I0112635

Earth and Life

EARTH & LIFE

A Four Billion Year Conversation

Andrew H. Knoll

PRINCETON UNIVERSITY PRESS PRINCETON & OXFORD

Published by Princeton University Press
41 William Street, Princeton, New Jersey 08540
99 Banbury Road, Oxford OX2 6JX

press.princeton.edu

GPSR Authorized Representative: Easy Access System Europe
Mustamäe tee 50, 10621 Tallinn, Estonia, gpsr.requests@easproject.com

ISBN 978-0-691-18223-0
ISBN (e-pub) 978-0-691-28731-7
ISBN (PDF) 978-0-691-27497-3
Library of Congress Control Number: 2025947007

British Library Cataloging-in-Publication Data is available

Editorial: Alison Kalett and Laura Lassen
Production Editorial: Jenny Wolkowicki
Text and jacket design: Chris Ferrante
Production: Jacqueline Poirier
Publicity: Matthew Taylor and Kate Farquhar-Thomson
Copyeditor: Charles J. Hagner

Jacket illustration by Eric Nyquist

This book has been composed in Oceanic Text, Geograph, and Antarctica

Printed in the United States of America

10 9 8 7 6 5 4 3 2 1

For Kirsten and Rob, with love always

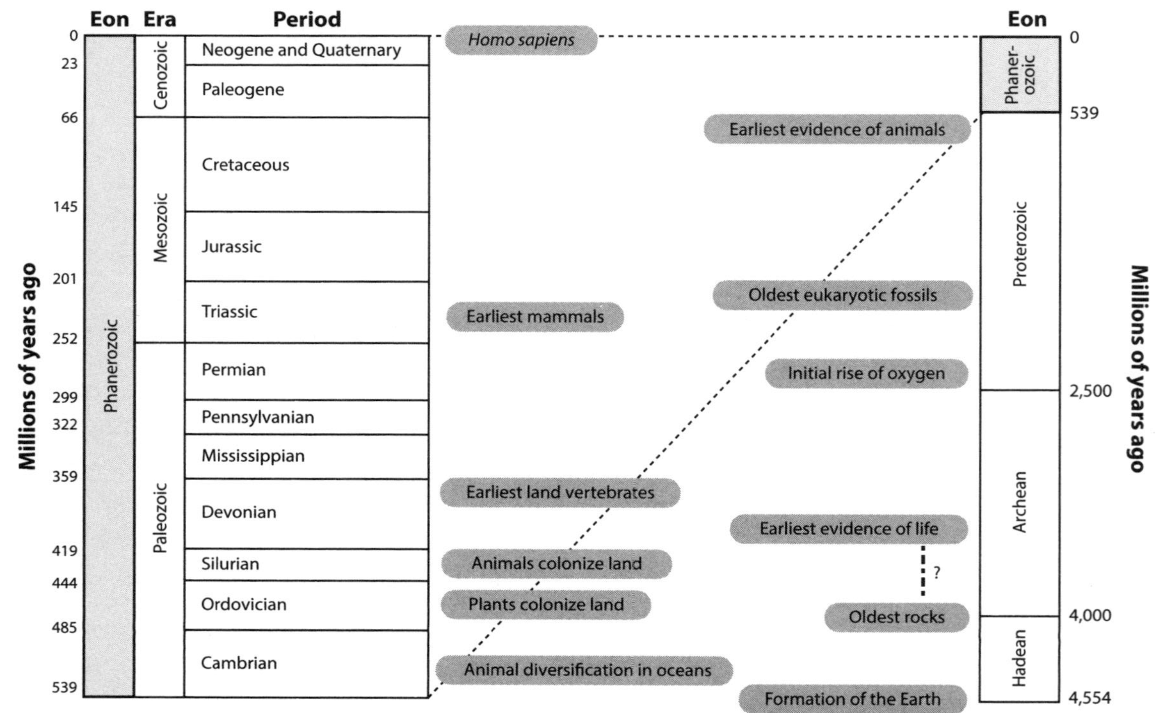
Eon
Era
Period
Millions of years ago
0
23
66
145
201
252
299
322
359
419
444
485
539
Phanerozoic
Cenozoic
Mesozoic
Paleozoic
Neogene and Quaternary
Paleogene
Cretaceous
Jurassic
Triassic
Permian
Pennsylvanian
Mississippian
Devonian
Silurian
Ordovician
Cambrian
Homo sapiens
Earliest mammals
Earliest land vertebrates
Animals colonize land
Plants colonize land
Animal diversification in oceans
Earliest evidence of animals
Oldest eukaryotic fossils
Initial rise of oxygen
Earliest evidence of life
?
Oldest rocks
Formation of the Earth
Eon
Phaner-
ozoic
Proterozoic
Archean
Hadean
0
539
2,500
4,000
4,554
Millions of years ago

CONTENTS

Earth and Life

CHAPTER 1

Discovering the Conversation

In the fall of 1973, I set out for college with high hopes and some trepidation, a bag of socks and underwear, and no idea what I wanted to do in life. In the small Pennsylvania Dutch town where I grew up, only a limited array of professions impinged on daily life, notably medicine, law, and engineering. I knew I didn't want to become a doctor or lawyer, and I was good at math, so I decided to study engineering. By the end of freshman year, I had one additional insight: I wasn't cut out for engineering, either.

At home, it was clear that "I don't know" wasn't a popular answer to the question of what I was doing in college, so when I returned to campus, I enrolled in five math and science courses and audited another, hoping that at least one of them might rub off. By some minor miracle, it worked. In fact, I was inspired by two courses, one in geology and the other in biology. As taught in those days, these courses seemed to describe separate universes, but sitting in my room one evening, it struck me that maybe they could be seen as two sides of a single coin. Maybe knowing something about biology would make me a better geologist, and, equally, knowledge of the Earth might help me to understand the life it supports. Fossils provided an obvious point where Earth and life intersect, but even as a sophomore,

I could dimly see that questions of how the Earth works—how, for example, carbon cycles through the biosphere—involve both physical and biological processes. I'm not sure that my teachers shared this nascent vision, but they were supportive and probably smiled inwardly when this naïve geology student showed up in a microbiology course. I, on the other hand, was hooked; the conversation between Earth and life would become the lodestar for my career in science.

I didn't know it at the time, but the way of thinking that took root in my undergraduate mind has a name: geobiology. Unlike physics and chemistry, which are fundamental approaches to matter and energy, geology is the study of an object: the Earth. When I was a student in the 1970s, applications of physics and chemistry to the study of our planet, called geophysics and geochemistry, respectively, were in the ascendancy, rapidly changing how scientists view the Earth and its history. In contrast, geobiology had yet to flower.

In the decades since then, geobiology—the study of how Earth and life interact and have done so through time—has expanded from a fledgling enterprise to a thriving discipline at the intersection of physical and biological sciences. Paleontology has benefitted enormously from geobiological insights, but so have studies of the environment—past, present, and future. Geobiology permeates discussions of life's great radiations and mass extinctions. It underpins research on Earth's environmental history. It illuminates many aspects of ecology and evolution and is critical to understanding twenty-first-century global change. Moreover, it informs astrobiological efforts to understand how life might be distributed throughout the universe. Disciplinary silos are so twentieth century; integrative approaches now fuel much of our most exciting and transformative science, including how Earth and life talk.

Despite its recent flowering, geobiology has deep roots. Snippets of geobiological thinking might be recognized retrospectively in the writings of da Vinci, Steno, and other Renaissance luminaries, but it was really in the eighteenth century that modern views of Earth and life began to take shape. For many centuries, Judeo-Christian tradition had held that our planet's history was short: An essentially modern-day Earth was formed during the first four days of Creation, to be populated with life on Day 5 and, on Day 6, Adam and Eve. James Ussher's 1650 calculation that it all began around nightfall on October 22, 4004 BC is famous for its precision, but in truth, others, from the Venerable Bede to Isaac Newton, had made similar estimates of the Earth's antiquity, all based on New Testament elaboration of the generations from Adam to Jesus.

Cracks in this young Earth chronology began to develop in the mid-eighteenth century, with the proposal by the great French naturalist Georges-Louis Leclerc, Comte de Buffon, that our planet must be much older than conventionally accepted; he estimated about 75,000 years based on experiments with the cooling of molten iron, which he envisioned to be important in Earth's development through time. Buffon also argued that this long planetary history is recorded in the rocks observed in cliffs and mountainsides.

By the century's end, the cracks had widened into emerging new views of the Earth. In a series of popular lectures and publications, a pioneering German mineralogist named Abraham Gottlob Werner hypothesized that the rock layers observable in nature reflect the sequential precipitation of different materials from an originally global ocean, beginning with granite and concluding with unconsolidated sediments. Called neptunism, after the primacy of precipitation from seawater, Werner's worldview envisioned a relatively old Earth (he reckoned a million years or so), with a discernible, deterministic history recorded in the rocks. The Earth around us took shape as minerals precipitated and the waters receded.

At about the same time, a Scottish naturalist named James Hutton was developing a different view of the Earth. A keen observer, Hutton recognized that our modern world is not invariant; hills and mountains are continually sculpted by erosion, while the sediments they generate inexorably fill in lakes and embayments. This understanding, however, laid bare a biological conundrum: If the environments that support life on land and in the sea are in a continual state of change, how can plants and animals, so clearly designed to fit their environs, persist in time? Hutton's solution was ingenious: As erosion wears down summits in one place, heat lifts new mountains upward somewhere else, maintaining the environments that species need to survive. Moreover, uplift and erosion, burial and lithification, continually cycle the materials of the Earth from mountain to sediments and back again. In a now famous 1788 essay, Hutton repeatedly used the term "habitability," and his view that life is sustained through time by planetary processes anticipates modern astrobiology by two centuries. Like Werner, then, Hutton saw history in rocks, but unlike Werner, he had no need to invoke processes no longer in play. In Hutton's "uniformitarian" view (from the hypothesized uniformity of process throughout Earth's history), our planet has been shaped and reshaped through time by processes we can observe today, leaving, in Hutton's famous phrase, "no vestige of a beginning, no prospect of an end."

Two centuries after the publication of Hutton's *Theory of the Earth* (1788), James Lovelock, who waits in the wings to enter this pocket history, reminded Earth scientists that Hutton actually likened the Earth to a superorganism and recommended what he called geophysiology as the proper approach to its investigation.

> *We are thus led to see a circulation of the matter in this globe, and a system of beautiful economy in the works of nature. This earth, like a body of an animal, is wasted at the same time that it is repaired. It has a state of growth and augmentation; it has another state, which is that of diminution and decay. This world is thus destroyed in one part, but it is renewed in another; and the operations by which this world is thus constantly renewed, are as evident to the scientific eye, as are those in which it is necessarily destroyed. (Hutton, 1788)*

Just as physiology regulates the properties of cells and tissues, keeping them within closely constrained limits (what biologists call homeostasis), the Earth is a self-regulating, geophysiological system, its endless changes maintaining the broad environmental constancy needed for life to persist. Organisms, however, really served only as a metaphor for Hutton. Life benefitted from this planetary regulation but played no major role in its maintenance. For several decades, the acolytes of Werner and Hutton debated the nature of Earth history, but in time, notably with the publication of Charles Lyell's influential *Principles of Geology* (1830–1833), the uniformitarian view carried the day.

One more crack in the view of an unchanging planet opened widely as the eighteenth century drew toward its close. Working at the Natural History Museum in Paris, the eminent anatomist Georges Cuvier established the fact of extinction—like the physical landscape, life was not constant in time. Many naturalists had observed fossils unlike any known living animals, but in an age of exploration, it seemed likely that while extant counterparts of those fossils might not reside in Europe, they would be discovered in the vast lands still to be explored. Moreover, to many contemporary thinkers, the very idea of extinction suggested flaws in the Creation, discouraging its consideration. Studying the bones of mammoths and mastodons, however, Cuvier recognized that while these remains were similar to the skeletons of elephants, they belonged to distinct species, species unknown in the modern world. As elephants were large and conspicuous, Cuvier rejected the idea that mammoths and mastodons would turn up in some territory yet to be explored, and so, in a pathbreaking 1796 paper, he argued that these cousins of living elephants are extinct.

In the years that followed, geologists established that not only did many fossils record extinct species, but fossil faunas appeared to change systematically through time. To Cuvier, this reflected catastrophes that episodically exterminated life, followed by renewed Creation to repopulate the world. Catastrophism provided one explanation for the emerging fossil record, but another, sharply different view was percolating among early-nineteenth-century thinkers: evolution.

Charles Darwin's *On the Origin of Species,* published in 1859, is generally regarded as the foundational document in evolutionary biology. Darwin proposed a compelling mechanism for evolutionary change, called natural selection, and so placed evolution at the center of debates about life and its history. He was not, however, the first to espouse evolutionary ideas. Beginning, once again, in the late eighteenth century, dozens of people had earlier argued the case. Many of these theses gained few adherents, requiring modern historians of science to achieve even minor recognition. But while evolutionary thinking was not widely embraced, it was fairly broadly known. For example, in his picaresque novel *Martin Chuzzlewit,* published 15 years before Darwin's magnum opus, Charles Dickens mentioned the idea that humans are descended from monkeys. His note has something of a "How 'bout them talking dogs" flavor to it, but Dickens clearly expected readers to know the reference. With Darwin and the subsequent birth of genetics, not only did evolution become biology's most important theory, it also opened new ways to consider how Earth and life talk.

Hutton, Werner, Lyell, and others established that Earth is a dynamic planet, its immensely long history recorded in rocks. In turn, Darwin and others argued that life is equally dynamic, with a long history recorded by fossils and explained by evolution. But it took one more visionary to introduce the idea that Earth and life interact in time and space. That person was Friedrich Wilhelm Heinrich Alexander von Humboldt. In the twenty-first century, Humboldt's is arguably not a household name, persisting in no small part through the distant echoes of place-names, from Humboldt, South Dakota ("A small town with a big heart," according to its website) and Humboldt County, California, with its eponymous university, to the Humboldt Glacier in Greenland and Berlin's Humboldt University. Two hundred years ago, however, Humboldt was among the most famous people in the world, a friend of Goethe and houseguest of Thomas Jefferson. Napoleon was cool toward his fellow Parisian, perhaps, it is said, out of jealousy for Humboldt's fame.

Humboldt was an intrepid explorer, as well as a deep thinker who excelled at seeing the connections among his various observations. From 1799 to 1804, he and his research partner Aimé Bonpland undertook the first scientific exploration of South America's interior, paddling along the Orinoco River and scaling volcanoes in the Andes. Prominent among Humboldt's many discoveries was the geographic distribution of plant species—in South America, the flora changed with altitude much as it did with latitude in Humboldt's European experience. From this, he concluded that the structure of plants and their geographic distribution are governed by their physical environment—

temperature, water, and soil. Importantly, and way ahead of his time, Humboldt reasoned that one couldn't, or at least shouldn't, study the flora and the environment in isolation. To understand either, one had to consider both together as a unified system. "Everything," he wrote, "is interrelated."

This integrated view of Earth and life shines through the great work of Humboldt's maturity, *Kosmos* (1845), but it eventually faded into the background as science grew more disciplinary and intellectual silos came to demarcate (and isolate) different fields. In her superb biography of Humboldt, Andrea Wulf captures its fate well: "Unlike Christopher Columbus or Isaac Newton, Humboldt did not discover a continent or a new law of physics. Humboldt was not known for a single fact or a discovery but for his worldview. His vision of nature has passed into our consciousness as if by osmosis. It is almost as though his ideas have become so manifest that the man behind them has disappeared." Humboldt, it is worth noting, also recognized that in time human activities might harm the natural world. "The most dangerous worldview is the worldview of those who have not viewed the world," he wrote, a thought worth taking to heart today.

So when did the outlines of modern geobiology begin to coalesce? The holistic perspective necessary to integrate Earth and life sciences began to (re)emerge early in the twentieth century, notably through the writings of the Russian mineralogist and geochemist Vladimir Ivanovich Vernadsky. His 1926 volume *The Biosphere* is commonly recognized as the opening salvo of modern geobiological thought, although it says something about the field's development that *The Biosphere* was not translated into English, even in abbreviated form, until 1986. Vernadsky's language is flowery, more metaphorical than analytical. He proclaims more than argues, and readers may find one bit nonsensical and another visionary, sometimes in the same paragraph. Without question, however, Vernadsky appreciated the interconnectedness of Earth and life, and his mineralogist's knowledge of chemical reactions helped him to recognize chemistry—what we now call geochemistry—as a primary basis for the connection.

In 1875, the Austrian geologist Eduard Suess coined the term *biosphere* as the envelope at the Earth's surface that contains and sustains life. In adapting this term, Vernadsky reframed it as the zone in which life and Earth interact. In particular, Vernadsky appreciated the role of metabolism in the conversation between our planet and the life it supports. Metabolism encompasses the ways in which organisms take in food and energy from their environment and use it for growth, reproduction, and work. Photosynthesis, for example, converts carbon from the environment into organic molecules, and respiration reverses the process, establishing a cycle that connects the physical and biological worlds. Indeed, Vernadsky voiced some specific appreciation for the role of bacteria (his student Sergei Winogradsky—he of the famous column found in many biology classrooms—would become one of the great pioneers of microbial ecology), but plants and animals dominate his discussion.

Vernadsky believed that the physical Earth has not changed appreciably through time and that life has always been a part of it. He accepted evolution but didn't consider it an important process in shaping the physical environment. He did, however, recognize that humans have changed the game, building on Pierre Teilhard de Chardin's slightly earlier concept of the noosphere (the biosphere in the age of reason). Anticipating the current concept of the Anthropocene (chapter 16), Vernadsky argued that much as the biosphere had transformed the geosphere, the noosphere was now transforming the biosphere.

The penultimate personality in this history is the Dutch scientist Lourens Baas Becking, whose 1934 book *Geobiology* not only gave the emerging discipline its name but clearly set out a modern view of how Earth and life talk. (Similar to Vernadsky's book, *Geobiology* was published in English translation only in 2016.)

Baas Becking was fortunate to study at Delft University during the golden age of microbiology that flowered there. He recognized that bacterial metabolism lay at the functional heart of geobiology, with microbes cycling nitrogen, sulfur, and other elements, as well as carbon. Establishing a lab at Stanford University, Baas Becking turned his attention to saline lakes, working out how microorganisms thrive in, and indeed help to regulate, these salty ecosystems. Although Bass Becking would return to the Netherlands and then move on to Australia, his early research on saline lakes gave him the experience and perspective needed to undertake his extended essay.

"This discourse is about this life, of and by the Earth," wrote Baas Becking, ". . . an attempt to describe the relationship between organisms and the Earth. The name 'geobiology' simply expresses this relationship. This new word does not attempt to describe a new field. It rather tries to unite phenomena that have thus far been known to the different areas of biology as much as possible under one viewpoint." He was too modest.

What followed was a detailed description of environmental factors that influence life and biological processes that help to shape the environment—sleeves-rolled-up empiricism in contrast to the philosophical musings of Vernadsky. Baas Becking's book contains what may be the first graphical depiction of a biogeochemical cycle, outlining how carbon moves through the biosphere. Baas Becking was more concerned with process than history, so a geobiological view of Earth and life through time was left for another generation. But his picture of how elements cycle through environments, and therefore how organisms and environments influence each other, is strikingly modern.

In the 1950s and 1960s, as geochemistry flowered as a way of examining the Earth, increasing attention started to be paid to element cycling, as advanced in seminal works by Robert Garrels, Robert Berner, Wallace Broecker, and others. At the same time, geobiology as an emerging field developed a time dimension, reorienting research on the history of life and environments. Dick Holland, one of my mentors in graduate school, pioneered geochemical research into our planet's environmental history, arguing that for the first two billion years of Earth history, the atmosphere and oceans contained little, if any, oxygen gas. At the same time, in a neighboring building, my principal thesis advisor, Elso Barghoorn, discovered fossils of ancient microorganisms which demonstrated that Earth has been a biological planet—a microbial planet—for most of its history. It was Preston Cloud, however, who linked long-term trends in environmental and biological history, effectively founding the subdiscipline of historical geobiology. A paleontologist by training, Pres recognized that Earth's tripartite oxygen history—none, a bit, and then a lot—coincided in time with a fossil record marked by bacteria, then bacteria plus protozoans and algae, and finally microbes plus plants and animals. Earth and life, Pres maintained, changed through time in coordinated fashion, with evolution both influencing and being influenced by the physical environment.

A central figure in this intellectual coming of age is James Lovelock, father of the Gaia Hypothesis. An independent inventor, notably of the electron capture detector, which allows scientists to measure exceedingly small amounts of chemical compounds in a sample, Lovelock was a frequent visitor to NASA's Jet Propulsion Laboratory at the time when Mars exploration was moving from aspiration to reality. Discussions at JPL inspired him to think about how one might detect the presence of life on another planet. While his colleagues focused on potential chemical signatures in Martian soils, Lovelock looked more broadly, his attention captured by the atmosphere. The thin atmosphere of Mars is decidedly distinct from that of Earth, and as Lovelock and his JPL colleague Dian Hitchcock discussed the problem, they came to the conclusion that the air we breathe on Earth must reflect life, and not simply the physical workings of the planet. Martian air, in contrast, showed no evidence of such biological influence. From this kernel, Lovelock developed a broader, more radical idea, which he christened the Gaia Hypothesis, after the personification of Earth and mother of all life in Greek mythology.

In Lovelock's view, Gaia is a complex entity involving Earth's biosphere, atmosphere, oceans, and soil, the totality constituting a feedback or cybernetic system that seeks and maintains an optimal physical and chemical environment for life on this planet. The maintenance of relatively constant conditions by active control may conveniently be described by the term borrowed from biology: homeostasis, already introduced as the physiological maintenance of stable conditions within cells or larger organisms. Here, then, is the essence of Gaia: Like Hutton's concept of geophysiology, Gaia envisioned a world kept habitable by interacting processes, but now with life, rather than Earth, in the driver's seat.

One day in the early 1970s, I happened to sit in on a fascinating conversation. Six people crowded around a small table in Harvard's Earth science building, as Jim Lovelock, not yet a household name, held forth on the relationship between life and environment. Organisms, Lovelock argued, were the architects of the physical world around them, actively maintaining Earth as a habitable planet. Dick Holland, already introduced as a game-changing geochemist, was having none of it. To Dick, at least at the time, life was more or less an epiphenomenon on a planet regulated by physical processes. As the only student at the table, I kept my mouth shut, but the discussion stayed with me.

Today, the end-member perspectives of Lovelock and Holland have largely been abandoned, replaced by a gamut of hypotheses from the broad intellectual space between them. It is far from clear what “optimal conditions for life” entail, as optimal conditions for sulfate-reducing bacteria lie far from those best suited for rabbits. Nor is it necessarily clear that natural selection will favor mutations that benefit the biosphere as a whole, rather than individual populations. And increasingly sophisticated research into our planet’s history strongly implicates the physical Earth in redox history, climate change, and mass extinctions. Yet Lovelock’s provocation unquestionably moved the dial. Life is not and never has been simply a passive presence on an active planet. Life influences the physical environment in ways both large and small, and with equal certainty, events in the physical Earth have shaped both ecology and evolution through time. It is not the predominance of biological and physical processes that matters, but, rather, their interactions. In a nutshell—geobiology.

CHAPTER 2

The Nature of the Conversation

Starting with C

In the summer, I will occasionally spend time meandering through the forests of central New Hampshire, continually amazed by the sheer dynamism of the woodlands' inhabitants. Woodpeckers drill tree trunks in pursuit of beetle larvae; gray squirrels dodge back and forth, ever vigilant against marauding foxes; owls perch on open branches, ready to pounce; while flowers bloom and then wilt, revealing fruits that will spread seeds across the forest floor. By comparison, the forests' physical surface of rock and soil appears to be passive, an unchanging platform for the dynamic biota. That, however, is misleading. Every atom in woodland animals, plants, fungi, and microbes was once part of the solid Earth, and it will be again. In forests, and in all other habitable environments on Earth, physical and biological processes interact continually, and those interactions shape the world around us.

If Earth and life talk, what do they say, and how do they say it? To converse, our planet and the life it supports must share a common language. We need letters, and these must assemble into meaningful words. Syntax, in turn, is necessary to assemble those words into meaningful messages. As already noted, the language of the Earth system is predominantly, although not entirely, one of chemistry. Its alphabet has only a small number of letters, most prominently C, H, O, N, P, and S—shorthand for the elements found most abundantly in living things. The words are molecules and minerals, ions and gases; and syntax is provided by physical and biological processes.

The logical place to start is with the letters, and here pride of place goes to C—carbon. Minerals are the building blocks of the solid Earth, and planetary scientist Shaunna Morrison and her colleagues have recognized more than 400 distinct minerals that contain carbon. Despite this diversity, carbon isn't particularly abundant in our planet; estimates for the bulk composition of Earth (table 2.1) suggest that carbon contributes only about 750 parts per million, by weight, and most of that is locked away deep inside our planet's core and mantle. (In parts-per-million parlance, 1 percent is equal to 10,000 parts per million—much greater than the proportional abundance of carbon in and on our planet.) In contrast, the cells in your body are about 18 percent carbon. As 60 percent or more of your cells' mass is actually water, carbon accounts for nearly half of the dry weight of your tissues and organs (table 2.2).

Why carbon? Why should an element that is so uncommon in our planet take center stage in your body, indeed, in all organisms? A major reason is carbon's chemical versatility; carbon atoms can join together to make a remarkable diversity of molecules, including the DNA that stores the cell's information, the proteins that endow our cells with structure and function, and the lipids that form

Table 2.1. Composition of the bulk Earth (PPM = parts per million)

Element	Abundance, by Weight (%)
Oxygen (O)	32.5
Iron (Fe)	26.2
Silicon (Si)	17.2
Magnesium (Mg)	15.4
Nickel (Ni)	1.6
Calcium (Ca)	1.6
Aluminum (Al)	1.5
Sulfur (S)	0.7
Sodium (Na)	0.25
Potassium (K)	0.02
Titanium (Ti)	0.07
Phosphorus (P)	1240 ppm
Carbon (C)	750 ppm

Source: Allègre and others, 1995.

membranes within and around each cell. Carbon atoms bind with other carbons, as well as with hydrogen, oxygen, and other elements, to form diverse chains and rings, each with distinct properties. Moreover, carbon is available to life as carbon dioxide gas (CO_2) in the air and as CO_2 and as carbonate (CO_3^{2-}) and bicarbonate (HCO_3^-) ions dissolved in fresh water and the oceans. These and many more carbon-bearing molecules dissolve readily in water, greatly enhancing their bioavailability. No other element combines such chemical versatility and availability at the Earth's surface. (A glance at the periodic table of the elements suggests

Table 2.2. The relative abundances of elements in human cells, by weight

Element	Composition (%)
Oxygen	65
Carbon	18
Hydrogen	9.5
Nitrogen	3.2
Calcium	1.5
Phosphorus	1.2
Potassium	0.4
Sulfur	0.2
Sodium	0.2
Chlorine	0.2
Magnesium	0.1
All other elements	< 0.1 each

Source: Helmenstine, 2018.

that silicon might work as well, but most of the silicon on Earth—and in our solar system—is inextricably bound to oxygen, leaving it unable to make the kinds of molecules found in living things.)

So carbon is concentrated in life, from bacteria to buffalo. Estimates of the total amount of carbon in Earth's biota run to about 750 gigatons (Gt; one gigaton equals one billion metric tons), a majority of it in plants. Plants actually cheat a bit here, as 90 percent or more of the carbon in the tree outside your home is found in wood, a nonliving tissue. If we consider only the living cells of plants, bacteria ascend to dominance in Earth's biological carbon reservoir. In either

Table 2.3. The distribution of biomass among major groups of organisms, in gigatons (1 Gt = one billion metric tons)

Group	Mass (Gt)
Plants	450
Bacteria	73
Fungi	12
Archaea	7
Protists*	4
Animals	2.5

Source: Bar-On and others, 2018.

* Protists include all eukaryotic organisms that are not animals, plants, or fungi.

case, animals make up only a small percentage of Earth's living biomass (table 2.3).

There's roughly the same amount of carbon in the atmosphere as there is in the biota, nearly all of it as carbon dioxide—perhaps 750 Gt in total (figure 2.1). While CO_2 forms only a small part of the air we breathe, it plays a major role in regulating Earth's climate—we'll unpack this in chapters to come. Soils, in turn, contain as much carbon as the biota and atmosphere combined: some 1,500 Gt, largely as the decaying remains of plants, animals, and microorganisms. Much more carbon can be found in the oceans. Little of this marine carbon resides in organisms (about 3 Gt), but as much as 38,000 Gt of carbon is dissolved in seawater, mostly as carbon dioxide and the two ions, carbonate (CO_3^{2-}) and bicarbonate (HCO_3^-). By far the largest surface reservoir of carbon, however, lies in sedimentary rocks: as much as 1,000,000 Gt. About 80 percent of this

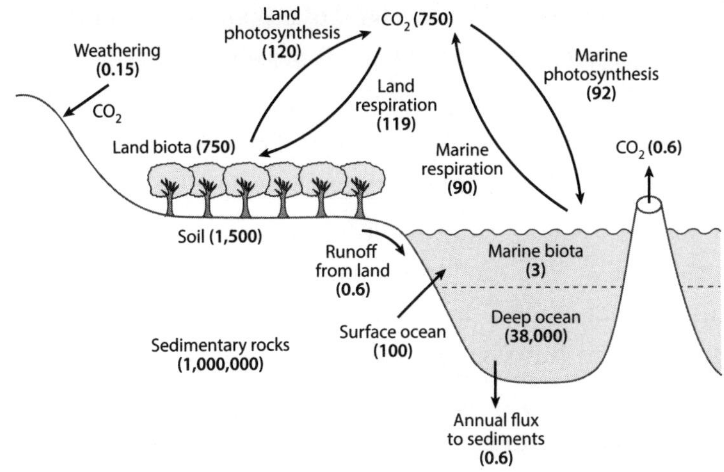

FIGURE 2.1. The global carbon cycle. Reservoir size and annual fluxes are presented in gigatons (Gt; 1 Gt = one billion metric tons). Sources and sinks may not balance, as reservoir and flux estimates come with significant uncertainties.

occurs in minerals, largely the calcium carbonate minerals that make up limestone ($CaCO_3$) and its mineralogical cousin dolomite [$CaMg(CaCO_3)$]. The rest consists of buried organic matter, only a tiny percentage of which is present as coal or petroleum. Actually, most of Earth's carbon resides in its interior. As noted above, the concentration of carbon in the crust, mantle, and core is low—perhaps 300–700 parts per million—but because the solid Earth makes up nearly all our planet's mass, the concentration accounts for most of Earth's total carbon inventory.

On various timescales, these reservoirs are in continuous communication (figure 2.1). For example, each year photosynthetic organisms remove some 210 Gt of C from the atmosphere, harnessing energy from sunlight to convert CO_2 into sugar, raw material for the diverse organic compounds needed for life; land plants and photosynthetic cells floating in the ocean contribute about equally to the total. That's a healthy percentage of all the carbon dioxide in the atmosphere, so if life is to persist over long intervals of time, there must be additional lines of communication between the atmosphere and other reservoirs—you, for example. As you read these pages, you exhale every few seconds, emitting CO_2 into the air around you. The CO_2 is generated within your cells by a process called aerobic respiration. When you eat, you introduce a variety of carbon-bearing compounds into your body. Some of the ingested carbon will be modified to form the molecules needed for growth, and some may be stored as fat—energy for another day. But much will react with oxygen gas to provide the energy you need to function; that's aerobic respiration. So, while photosynthesis takes CO_2 and water and, powered by light energy, generates organic molecules and oxygen gas, aerobic respiration essentially runs the reaction in reverse, reacting oxygen gas and organic compounds and releasing CO_2 and water in the process.

Together, then, photosynthesis and aerobic respiration form a cycle, continually shuttling carbon from the environment into cells and back again—a prime example of Earth and life in conversation. Nearly all plants, animals, fungi, and eukaryotic microorganisms respire aerobically, as do many bacteria and their microbial cousins, the archaea. In aggregate, all of this respiration almost balances photosynthetic CO_2 removal from the atmosphere, sustaining an atmosphere that can support life. Almost, but not quite. A percent or so of the organic matter generated by photosynthesis escapes respiration and related processes to accumulate in sediments. Through time, this leak in the biological carbon cycle builds and maintains the reservoir of organic matter in sedimentary rocks.

Before delving into the fate of the organic carbon that accumulates in sediments, it is worth pausing to look more closely at the carbon cycle as we've described it so far. Figure 2.1 actually contains two element cycles. As C is cycled between CO_2 and organic matter, oxygen, or O, cycles between water (H_2O) and oxygen gas (O_2). What happens if O_2 is absent? Such environments may be unfamiliar, but they are widespread on the modern Earth: in the black mud of swamps and marshes, in the aptly named dead zones and oxygen minimum zones of the oceans, and in sediments where ambient oxygen has been depleted by aerobic respiration. Moreover, as we'll explore in a subsequent chapter, life was present for more than a billion years before oxygen began to accumulate in the atmosphere and surface ocean, and the deep waters of the sea have been consistently oxygenated for only the most recent 10 percent (or less!) of life's history.

It turns out that carbon cycles through oxygen-poor environments just as it does in our familiar O_2-rich world. The distinction lies in which organisms do the cycling and what chemical compounds they deploy. Where oxygen is present, the cycle includes well-known processes carried out by familiar organisms. Plants take carbon dioxide from the environment and convert it into organic matter; animals (and plants as well) respire organic matter, returning CO_2 to the air or water that surrounds them.

Fungi, protozoans, algae, and many bacteria and archaea participate as well.

In anoxic—that is, oxygen-depleted—environments, eukaryotic organisms generally play only a limited role in cycling carbon. Bacteria and archaea do most of the work. In terms of size and shape, these tiny microorganisms display only limited diversity, but their metabolic diversity is dazzling. Among bacteria, eight different groups are known to photosynthesize, and only one of them—the cyanobacteria—uses water as a source of electrons needed to convert carbon dioxide into sugar, emitting O_2 in the process. The others obtain electrons from compounds like the hydrogen sulfide (H_2S), hydrogen gas (H_2), or reduced iron (Fe^{2+}) found in anoxic waters, generating molecules like sulfate (SO_4^{2+}) and oxidized iron (Fe^{3+})—but not O_2—as by-products. Similarly, while many bacteria and archaea are capable of aerobic respiration, others respire using oxidants other than O_2, including sulfate (SO_4^{2-}), nitrate (NO_3^-), ferric iron (Fe^{3+}), or even oxidized forms of arsenic (AsO_4^{3-}). Collectively, these microorganisms can cycle carbon in the absence of O_2, connecting photosynthetic microbes that do not generate oxygen with respirers that don't use it. Note that just as C cycling is linked to the cycling of water and oxygen gas in our familiar world, the C cycle in oxygen-poor environments is closely tied to other cycles, including those of sulfur and iron (chapter 4). And there's more: Bacteria and archaea exhibit a variety of fermentation reactions, the breakdown of sugars and other complex organic compounds into two smaller molecules in the absence of oxygen gas, and some gain energy by a process called chemosynthesis, relying on chemical reactions, rather than sunlight, to generate the energy needed to fix CO_2 into sugar.

Bacterial metabolisms have played a major role in the carbon cycle through Earth history, and they continue to do so today. And, in addition to cycling carbon formed or transported to oxygen-poor environments, these microbes find importance in human cultures around the world. For example, bacteria that ferment sugar to acid give us pickles, kimchi, sauerkraut, vinegar, and yogurt. Yeasts (the one group of eukaryotes that really excels at fermentation) get credit for beer, wine, and leavened bread.

Ecosystems are complicated machines with many moving parts, but at heart they populate a slightly leaky cycle that moves carbon from the environment to organisms and back again, connecting life in all its diversity to the planet on which it resides.

Now we can return to an issue raised earlier. If some of the carbon incorporated into organisms evades respiration and other metabolisms that return it to the environment, and instead accumulates in sediments, why doesn't this imbalance eventually drain CO_2 from the atmosphere? Once again, we must look to additional processes, this time physical processes at work in the Earth system.

The biological carbon cycle works on short timescales, scales that humans can observe and measure. When we integrate life into the larger geologic carbon cycle, however, the time frame expands dramatically, and quantification can be challenging. Like organisms, physical processes can either add carbon to the surface environment or remove it.

Removal is subtle, as chemical reactions slowly but inexorably wear down mountains. In contrast, physical processes that add carbon to the surface environment can engender fear and awe: volcanoes in action.

The jagged peaks of the Alps or Rocky Mountains project a stately air of permanence, stony sentinels seemingly chiseled for all time. In truth, however, weathering and erosion are continually wearing away these mountains, shaping and reshaping both peaks and the valleys between them. Weathering refers to the in situ alteration of rocks and soil by interactions with the ambient environment; erosion is the breakdown and removal of materials, transported by water, wind, or ice. Chemical weathering is particularly important to our story. Carbon dioxide in the atmosphere reacts with water to form carbonic acid (H_2CO_3), a weak acid that slowly dissolves exposed rocks. Ions freed by dissolution are whisked away by rain and rivers, bound, eventually, for the ocean. As carbonic acid does its work, its carbon is converted to bicarbonate ions (HCO_3^-) that also head toward the sea. Once these bicarbonate ions and calcium ions weathered from the rocks reach the ocean, they combine to precipitate as calcium carbonate ($CaCO_3$) minerals, limestone that accumulates on the seafloor. If you follow the carbon, then, you'll see that weathering and erosion transfer C from CO_2 in the atmosphere to limestone on the seafloor (or lake floor). This, of course, doesn't solve the imbalance of the biological carbon cycle; it makes it worse.

Enter volcanoes. No other physical feature stirs humans quite like volcanoes, their incandescent lavas flowing over the surrounding landscape while ash and lava bombs explode high into the atmosphere. It is no wonder that Fuji and Kilimanjaro have long been venerated as sacred by people on surrounding plains. At the same time, the destructive ferocity of Vesuvius is still recounted to schoolchildren, nearly 2,000 years after its catastrophic eruption. Even as I began to draft this chapter, volcanoes in the Canary Islands and Iceland spread lava over crops and towns.

As they erupt, volcanoes inject large amounts of gas into the environment, mostly water vapor, sulfur dioxide, and carbon dioxide. Emissions from dormant volcanoes, deep-seated fault systems, and mid-ocean ridges, where oceanic crust originates, contribute just as much. In total, these sources add an estimated 0.3–0.6 Gt of CO_2 to the atmosphere each year. This is much less than annual amounts of photosynthesis and respiration but is closer to the biological carbon cycle's net flux of carbon from air to sediments and the yearly removal of CO_2 by weathering. Other important players include CO_2 emitted during the metamorphism of limestones, the dissolution of carbonate rocks during weathering, and the oxidation of ancient organic matter in sedimentary rocks exposed to weathering and erosion.

Where does the carbon in volcanoes come from? Lava originates within Earth's mantle, as molten materials rise to the surface. It stands to reason, then, that the mantle must be the source of volcanic CO_2 as well. To gain perspective on this, it's helpful to view volcanoes within the framework of plate tectonics, Earth's internal heat engine that shapes geology, geography, and much more (see chapter 7). Our planet's surface is partitioned into a number of coherent units called plates, seven big ones and a number of smaller pieces. Each consists of crust and a thin layer of mantle beneath it, its aerial extent marked by boundaries that can be divergent, where new crust is formed; convergent, where one plate sinks beneath another, a process called subduction that returns crust and overlying sediments to the mantle; or transform, where two plates slide past each other—for example, the San Andreas Fault in California. Convergent boundaries are commonly marked by linear arrays of volcanoes, as seen, for example, in Japan or the Aleutian Islands. These volcanoes reflect the partial melting of crust and overlying sediments as they sink downward into the mantle. As a result, much of the CO_2 emanating from such volcanoes comes from carbon in the sinking crust itself and both carbonate and organic carbon in overlying sediments. Other volcanoes, however, lie far from plate boundaries, their lavas formed from molten material that wells up from deep within the mantle; the Hawaiian Islands illustrate the point. The carbon released by these volcanoes originates deep within the mantle. The same goes for gases that emanate from divergent boundaries such as the mid-ocean ridge system that runs down the center of the Atlantic Ocean.

That's not to say, however, that the carbon dioxide spewing from such features is primordial, trapped in the mantle since Earth was young. Some may be, but most reflects the admixture of surface carbon mixed downward into the mantle by tectonic processes. Evidence for this can be found in the mantle's best-known species of carbon—diamonds. Diamonds are pure carbon, formed at the high pressures and temperatures of Earth's interior. Tellingly, some diamonds have tiny inclusions of material that link them to a sinking slab of seafloor. Moreover, in these diamonds, the ratio of carbon's two stable isotopes, ^{12}C and ^{13}C, strongly suggests derivation from biologically formed organic matter, another indication that surface carbon has been mixed downward into the mantle through time. The mixing is uneven, so the amount of carbon in the mantle varies markedly from place to place, as inferred from measurements of gas emissions from volcanoes and oceanic ridges around the world. Much remains to be learned about Earth's deep carbon chemistry, but the key point is clear: Physical processes transport carbon from Earth's interior to the surface and back again, providing a long-term geological carbon cycle that links to biology's cycling on shorter time scales.

C, then, shapes much of the conversation between Earth and life. Earth's carbon was forged billions of years ago in the interior of ancient stars. Released into space as parental stars died, sometimes violently, carbon atoms were swept up with other materials as Earth took shape around our nascent Sun. From there, carbon's journey has been peripatetic: degassing from the mantle into the atmosphere; partitioning into different reservoirs in minerals, cells, gases, and ions dissolved in the sea; and shifting repeatedly from one reservoir to another via processes that run from metabolism to plate tectonics.

Viewed as a whole, Earth's carbon cycle may seem forbiddingly complex, an intricate meshwork of gears large and small, like Charlie Chaplin's hilariously complicated factory floor in *Modern Times*. But don't let the details obscure the broader picture. Diverse organisms rapidly cycle carbon from the environment into cells and back again. The cycle is leaky, slowly but surely moving carbon to sediments, a geologic reservoir. And like the metabolisms that populate Earth's biological carbon cycle, physical processes can either return geological carbon to the atmosphere or sequester more of it into sediments or, eventually, the mantle. Over long timescales, this continual movement of carbon through the Earth system keeps our planet more or less in balance, but the balance isn't perfect, and it can be disrupted by large transient events. Giant volcanoes that drove mass extinctions in the past furnish one example. Technological humans provide another. Through it all, however, carbon has woven life and Earth together since our planet was young, sustaining a habitable environment, and life itself, over billions of years.

CHAPTER 3

Expanding the Alphabet

N and P

Photosynthesis is arguably life's greatest invention, the one that freed early organisms from local environments where chemical compounds could be exploited for energy, enabling life to spread across the globe. Certainly, photosynthesis made life a key component of Earth's carbon cycle. In chapter 2, we noted that each year photosynthetic organisms remove some 210 Gt of carbon from the environment, but what controls this? What determines the amount of photosynthetic carbon fixation in a forest, in grasslands, or in the central gyres of oceans? Why don't plants and algae remove more carbon? Or less?

A clue comes from agriculture. Farmers have long known that when they spread fertilizer across a field, crop growth improves. Something in the fertilizer increases photosynthetic output in local plants. That something is nutrients—more specifically, two additional letters of the alphabet in life's conversation with Earth: N and P. Nitrogen and phosphorus. Why should N and P set limits on photosynthetic carbon fixation? The simple reason is that while organisms need carbon to grow and function, they also need other elements—none more important than N and P. Proteins contain N; DNA contains N and P, as does its biochemical cousin RNA; membranes contain P; and the energy currency of the cell is adenosine triphosphate, or ATP, again rich in P. Your cells may have 18 percent C, but they also clock in at about 3 percent N and a bit more than 1 percent P (table 2.2). You get the nitrogen and phosphorus needed for growth from the same place you obtain carbon—the foods you eat. If your dinner included beef or salmon, those animals also obtained C, N, and P from their diet. Ultimately, all organisms that consume others depend on primary producers, the (mostly) photosynthetic organisms that obtain their C, N, and P from the physical environment.

Plants are bathed in nitrogen, as are you and every other organism at the Earth's surface. Nitrogen gas (N_2) makes up about 78 percent of the air we breathe, so it might seem a simple matter for photosynthetic organisms to obtain the nitrogen needed for growth. There is, however, a problem: By themselves, plants and algae can't incorporate N_2 into biological molecules. To synthesize molecules like proteins and nucleic acids, plants depend on other chemical forms of nitrogen found in soils and aquatic environments, mainly ammonia (NH_3) or ammonium ion (NH_4^+) and nitrate (NO_3^-), which gets reduced to ammonia inside cells.

This, as ever, prompts another question: Where do those bioavailable nitrogen compounds come from? A major source is the decay of dead organisms. A variety of fungi and bacteria attack N-bearing biomolecules in soils, lakes, and the sea, releasing ammonia in the process. Aptly, biologists call this ammonification. Microbes also free up bioavailable ammonia from nitrogenous compounds in excreted wastes; urine, for example, is a rich source of nitrogen. The ammonia formed from decaying cells and wastes can be incorporated by primary producers, helping to cycle nitrogen between life and environment.

It turns out, however, that primary producers aren't the only organisms interested in that ammonia. A diversity of bacteria and archaea use ammonia in energy metabolism. Nitrification is a form of chemoautotrophy in which microbes react ammonia with oxygen gas to provide the energy needed to fix carbon dioxide into organic molecules. Nitrification occurs in two steps, with the oxidation of ammonia to nitrite (NO_2^-) followed by the further oxidation of nitrite to nitrate, generally by two distinct but interacting groups of microorganisms. Nitrifying chemoautotrophs, thus, compete with plants and other photoautotrophs for ammonia in the environment. Many photosynthetic organisms can take up the nitrate generated by nitrification for use in biosynthesis, but nitrification also introduces a "leak" into the nitrogen cycle, as a small percentage of the N in ammonia ends up as nitrous oxide (N_2O), known colloquially as laughing gas for its euphoric effect on humans. Much of this gas makes its way into the atmosphere, where it functions as greenhouse gas. N_2O is a minor constituent of the air we breathe, but its greenhouse effect on a per-molecule basis is about 300 times that of carbon dioxide. (Interestingly, this is not because N_2O absorbs radiation efficiently, but because it absorbs at wavelengths where CO_2 and water vapor do not absorb well.) As fertilizer use has increased, so has the atmospheric concentration of N_2O (chapter 16).

Yet another biological process puts a strong dent into the bioavailable nitrogen pool, as microbes once again compete with primary producers, this time for nitrate in soils or water. In chapter 2, we noted that where oxygen gas is limited or absent, some bacteria and archaea can respire using other molecules. One of these alternative oxidants is nitrate; the bacteria (and some archaea) that respire in this way are called denitrifiers. Much as we use O_2 to oxidize organic compounds during aerobic respiration, denitrifying microbes rely on nitrate (or nitrite) to respire biomolecules anaerobically, providing the energy needed for growth and reproduction. Nitrogen gas is produced as a by-product, transferring N from the bioavailable pool to the large atmospheric reservoir of N_2 that is inaccessible for growth. Incomplete denitrification also adds to the environmental abundance of nitrous oxide. Microbiologists estimate that each year denitrification removes more than 200 billion kg of N from the bioavailable nitrogen pool of the oceans.

One other process, only recently discovered, also depletes Earth's pool of bioavailable nitrogen. In the 1990s, scientists reported an unexpected chemoautotrophic pathway in which bacteria react ammonia and nitrite together, with the N in both molecules released to the environment as N_2. Called anaerobic ammonia oxidation, or anammox, this process is now known to be common in the oceans, accounting for an estimated 30–50 percent of bioavailable nitrogen removal from marine waters.

So far, the parts of the biological nitrogen cycle we've introduced result in a net movement of N from bioavailable pools in soil and water to unavailable N_2 in the atmosphere. Clearly, then, the long-term maintenance of life requires an additional process, one that transfers N from air to a bioavailable form—and so to life. That process is nitrogen fixation. Next to photosynthesis, nitrogen fixation may be the most important metabolism ever evolved on Earth, enabling abundant primary production to be sustained throughout our planet's history. And once again, nitrogen fixation is carried out by bacteria and archaea (a few of which live in close association with plants or algal cells)—more evidence that plants, animals, and other eukaryotic organisms live in a world shaped largely by microorganisms.

In nitrogen fixation, microbes convert nitrogen gas to ammonia. This reaction requires a great deal of energy, but it provides bioavailable nitrogen for the entire biosphere. In fairness, several nonbiological reactions can also convert N_2 to bioavailable N, including lightning that arcs through the atmosphere. But, quantitatively, biological nitrogen fixation is the key process in maintaining the nitrogen cycle that supports all life on Earth. (Humans fix a great deal of nitrogen by means of industrial processes; discussion of this, and its environmental consequences, is taken up in chapter 16.) Nonbiological reactions can also parallel the reactions involved in nitrification and denitrification, but, again, it is microbes that contribute most to Earth's nitrogen cycle.

At some point in your education, you probably learned that farmers commonly grow soybeans in crop rotation because of their capacity to enrich soils with bioavailable nitrogen. And so they do, but in a very particular way: Soybeans can't fix nitrogen any better than you can, but nodules on their roots contain populations of bacteria that can and do fix N_2. This association is a prime example of symbiosis, a close relationship between two species in which each partner contributes to the welfare of the other. The nodules on soybean roots provide a local oxygen-free environment where N_2-fixing bacteria can thrive, as well as nutrients for their growth. In turn, the bacteria leak bioavailable nitrogen molecules that the soybeans use to grow. Clover, alfalfa, and beans also harbor nitrogen-fixing symbionts, as do alders and a number of trees in tropical forests. In the oceans, some planktonic algae also maintain nitrogen-fixing bacterial symbionts within their cells, again contributing in important ways to the nutrient status of the environments they inhabit. Remarkably, in 2024, biologists reported that nitrogen-fixing bacterial symbionts in the unicellular marine alga *Braarudosphaera bigelowii* have become sufficiently tightly integrated into their host cells to qualify as novel organelles, much as the bacterial predecessors of mitochondria and chloroplasts did more than a billion years ago.

So, if we back off and view the biological nitrogen cycle as it emerges from the preceding discussion, it has much in common with the carbon cycle (figure 3.1). Complementing this, table 3.1 shows how nitrogen is distributed on and within our planet. A large proportion of Earth's nitrogen resides in the atmosphere—some four quadrillion tons—but even more is locked deep in the crust and mantle, perhaps twice as much as the atmospheric reservoir (table 3.1). Although this deep-seated geological reservoir is large, it interacts with the surface only slowly, over long timescales.

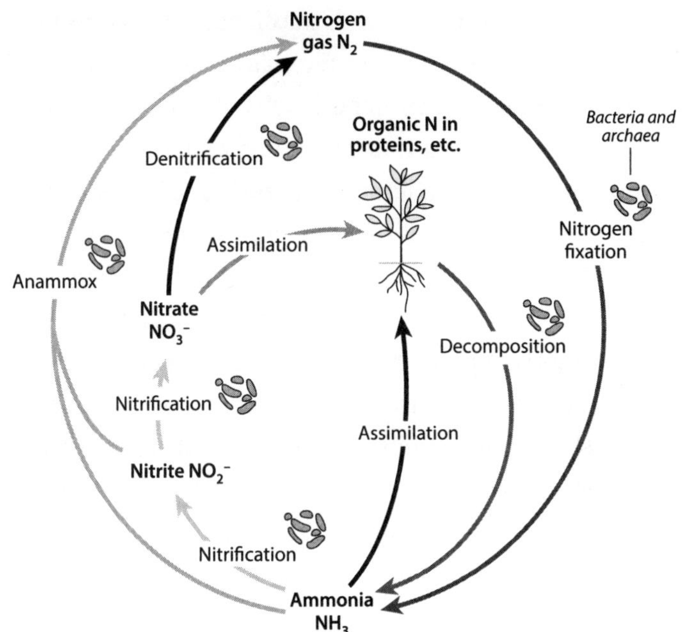

FIGURE 3.1. The biological nitrogen cycle, populated largely by bacteria and archaea. Every nitrogen atom in your body has passed through this cycle innumerable times.

Table 3.1. Nitrogen abundance in Earth and life

Reservoir	Mass N (g)
Atmosphere	3.9×10^{21}
Biota	4.3×10^{15}
Soil	1×10^{17}
Surface ocean	6×10^{16}
Deep ocean	6×10^{17}
Crust	1.1×10^{21}
Mantle	8×10^{21}

Source: Palya and others, 2021.

Interestingly, careful research published in 2020 suggests that much of the nitrogen in Earth's interior was once at the surface, later to be returned to the interior via the same subduction processes that transport carbon back to the mantle. There is much less nitrogen in soils and seawater, perhaps a thousandth of the atmospheric abundance, and the amount of nitrogen in organisms is even smaller, about a millionth of the amount in air. Earth can sustain a large and complex biomass because microbes convert nitrogen gas in the atmosphere to molecules that primary producers can use, while a variety of bacteria and archaea use bioavailable nitrogen in metabolism, eventually returning it to the large atmospheric N_2 pool.

Two final words about nitrogen: First, it is worth reemphasizing that while plants and animals participate in Earth's nitrogen cycle, it is bacteria and archaea that do the heavy lifting, employing metabolic pathways that simply do not occur in plants, animals, or other eukaryotic organisms. And second, the nitrogen cycle interacts strongly with the carbon cycle. Indeed, the environmental availability of nitrogen, mediated by N-metabolizing microbes, plays a key role in determining rates of photosynthesis, both locally and globally. In the absence of microbial nitrogen fixation, Earth could sustain only a limited biota. We'll return to this issue, but first, let's introduce one more key letter in the Earth-life alphabet: P.

If N is one of the key alphabetical pillars that support ecosystems, P—phosphorus—is the other. By weight, phosphorus makes up a bit more than 1 percent of your cells, mainly in the structural backbone of nucleic acids, in membranes, and in ATP. In fact, however, your body contains much more phosphorus, as your bones are hardened by calcium phosphate minerals. Unlike carbon and nitrogen, phosphorus rarely participates in the redox reactions that drive metabolism. Also, unlike C and N, P does not have a significant presence in air. By far, most of Earth's phosphorus is found in rocks and sediments, as apatite and related calcium phosphate minerals. Phosphorus makes up only about one-fifth of a percent of the solid Earth (table 2.1), but this is the source of the phosphate released into the environment as phosphate ions (PO_4^{3-}) via chemical weathering and erosion, eventually to be taken up by organisms. As was true of nitrogen, you and other animals obtain the P needed for growth and reproduction from the food you eat, whereas primary producers take up phosphate from their surroundings.

The P cycle begins with chemical weathering and erosion, which free up phosphate ions from minerals for incorporation by primary producers. On land, phosphate ions in soil are taken up by plants and photosynthetic microbes, eventually to be incorporated into consumers. Decomposition of dead organisms returns P to the environment, where it may be taken up once more by primary producers or may be carried by rivers to the ocean. In the sea, dissolved phosphate is taken up by primary producers in sunlit surface waters, to cycle rapidly among algae, heterotrophic organisms, and seawater. Viruses play an important part in this cycling; it is estimated that every day viruses cause 20–40 percent of all phytoplankton cells in the oceans to rupture. This kills the phytoplankton cells, freeing virus particles to infect more cells and returning nutrients into the surrounding seawater. Some organic matter eventually sinks downward to the deep sea as fecal pellets and dead cells, commonly aggregated into fluffy particles called marine snow. In the vastness of the ocean's depths, bacteria, protists, and animals live by respiring this material, again releasing nutrients in the process. Phosphate liberated in the deep sea returns to the surface via rising water masses, a process called upwelling. Indeed, in present-day continental shelf waters, most of the phosphate available for primary producers in surface waters is supplied by upwelling; weathering and erosion of adjacent landmasses account for only a few percent of the total.

Like the biological carbon cycle, the biological P cycle is leaky, and phosphate is eventually returned to sediments in skeletons, buried organic molecules, and phosphate minerals formed within the sediment. Phosphate in sediments and the sedimentary rocks they become can be sequestered for many millions of years, reintegrating with the biological P cycle only when mountain building thrusts the rocks above sea level, where they are again subject to weathering and erosion (figure 3.2).

By now, it should be clear that C, N, and P all cycle through the Earth system via both rapid biological processes and physical mechanisms that work on geologic timescales. And, as every organism requires all three elements, their cycles must be interlinked. So, to go back to an issue raised earlier, which—if any—of these elements limits photosyn-

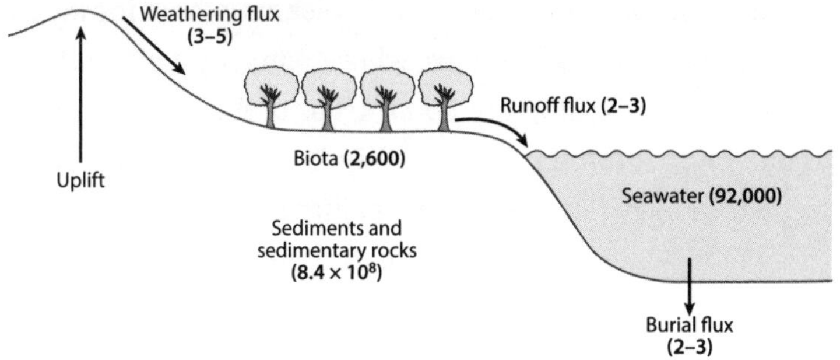

FIGURE 3.2. The global phosphorus cycle. Reservoirs are in teragrams P (Tg; 1 Tg = one million metric tons), and fluxes are in Tg P per year.

thesis? Logically, primary production has to be limited by the element present in the lowest abundance relative to the needs of the primary producer. Ecologists refer to this concept as Liebig's Law of the Minimum, after Justus von Liebig, a nineteenth-century German chemist, who popularized the idea. Materials that limit primary production are commonly called limiting nutrients.

Many scientists hold that, on geologic timescales, phosphorus limits primary production in ecosystems around the world. The logic underpinning this assertion is that while P availability is set largely by physical processes, nitrogen-fixing microbes can make up any shortfall in bioavailable N. That logic is reasonable, but in present-day ecosystems at least, empirical evidence tells a more complicated story. In fields and forests, the addition of N and P generally increases photosynthetic yield, but so does the addition of N alone. In contrast, treatment with only P commonly has little influence on photosynthetic output. This suggests that ecosystems on land are commonly limited by nitrogen availability. Nitrogen limitation also seems to hold for much of Earth's low-latitude oceans, where nutrient availability in general is low. Regionally, some areas of the ocean appear to be either P-limited or jointly limited by N and P, but in some regions where upwelling brings abundant nutrients to surface waters, neither P nor N seems to limit primary production. Rather, rates of photosynthesis can be limited by iron, or even cobalt.

Iron? Cobalt? An average adult male human contains about 4 g of iron (a bit more than 0.1 oz; because of their reproductive biology, women have somewhat less), and its abundance in plants and phytoplankton is similarly minute. Iron, however, plays a number of critical roles in cells—for example, in the hemoglobin that transports oxygen through our bodies and key enzymes required for photosynthesis. In the modern oceans, iron is scarce, because in oxygenated seawater, it is essentially insoluble, precipitating rapidly as iron hydroxides, or rust. Thus, iron can limit photosynthesis, not because photosynthetic organisms need a lot of it, but because it is generally scarce in seawater. Cobalt is even scarcer in cells; your body contains just a few milligrams, but it is needed for vitamin B12 and neurotransmitters essential for nervous-system function. Like iron, cobalt is generally scarce in soils and seawater, so much so that although organisms need only tiny amounts to function, its scarcity can be limiting for primary producers. The main point is that because of nutrient limitation, the cycles of N, P—as well as iron and cobalt—strongly influence the biological cycling of carbon. And they have done so throughout Earth history.

To get a picture of how the varying abundances of nutrients play out in the oceans, it is helpful to look at a map—not just any map, but a map of primary production in the oceans, made possible by satellite measurements of chlorophyll abundance (figure 3.3). Because chlorophyll reflects specific wavelengths of light, observations from space enable scientists to estimate its abundance and, by extension, primary productivity. In the subtropical to tropical central gyres of oceans, nutrient levels are exceedingly low and so, correspondingly, are rates of photosynthesis. Small size is useful in these environments, because smaller cells have a larger ratio of surface area to cell volume, facilitating the uptake of nutrients needed for growth and reproduction. In consequence, tiny primary producers, mostly cyanobacteria, dominate

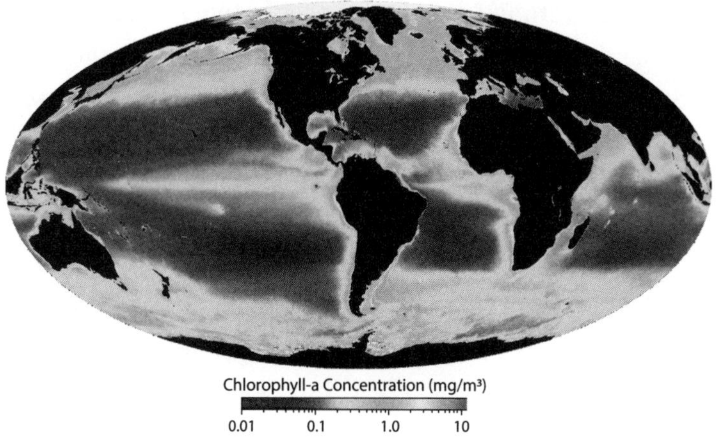

FIGURE 3.3. The geographic distribution of primary production in the oceans, based on satellite measurements of chlorophyll content in surface waters. The lightest areas, largely along coastlines, denote places with high rates of primary production; the darkest regions, largely in the central gyres of oceans, show where primary production is low.

primary production. Although the rate of photosynthesis per cell is low in these environments, their extent is vast, so the cyanobacteria they harbor account for as much as a quarter of all primary production in the sea.

In coastal environments that receive nutrient input from land and where upwelling rates are high, nutrients are more abundant, driving higher rates of photosynthesis. In these environments, eukaryotic phytoplankton become more important, green algae at first and then, as nutrient availability continues to increase, larger phytoplankton cells, such as diatoms. All in all, then, the distribution of nutrients, governed largely by physical processes, determines the spatial distribution of both photosynthetic rates and phytoplankton composition. And the distributions of different primary producers vary in time as well. For example, in year-long studies of primary productivity in the Atlantic Ocean near Bermuda, cyanobacteria dominated in the summer, when upwelling of nutrients was limited, but green algae came to the fore in the winter, when more nutrients moved upward into sunlit waters from below. The key point is that the physical characteristics of primary producers—perhaps especially cell size—influence their responses to nutrient levels in time and space. Predators play a role as well, limiting the abundance of favored prey and therefore facilitating the growth of other phytoplankton species. And increasingly, it is becoming clear that viruses, ubiquitous in seawater, can similarly influence the abundance and competitive abilities of phytoplankton populations.

In 1934, Alfred Redfield published a remarkable observation that has reverberated through the world of marine ecology ever since that time. Despite the evident heterogeneity in nutrient levels and primary-producer identity in the oceans, when Redfield measured the relative abundances of C, N, and P in seawater samples drawn from throughout the world's oceans, he found that the molar ratio (that is, the ratio of atoms) of C to N to P nearly always fell in the vicinity of 106:16:1. Interestingly, the C:N:P of marine phytoplankton tends to exhibit a similar relationship.

Marine scientists have been debating explanations for Redfield's observation for decades, with no end in sight. At some level, the Redfield ratio must reflect the biochemical composition of cells and the ways that the C, N, and P cycles feed back onto the environment. That said, increasing measurements show that C:N:P can vary substantially from species to species and from place to place, with light, temperature, growth rate, and other parameters influencing observed C:N:P. Curiously, the prediction that rates of primary production in some parts of the oceans should decline as twenty-first-century global change decreases nutrient availability has not come to pass. It appears that some combination of changing phytoplankton composition and adaptable physiology within species has resulted in a lower demand for N and P, maintaining rates of carbon uptake.

In land plants, the ratio of N to P is comparable to that of marine phytoplankton, averaging about 13:1, albeit with strong variations among species. N:P in soils is broadly similar. Relative to phytoplankton, however, plants tend to have significantly more C relative to N and P, in no small part because in land plants, water is transported through the body via tubes made of thick, porous walls (lots of C, no N or P) that no longer have internal cell contents. Particularly in large trees, most of the plant's mass consists of wood, the ultimate expression of cytoplasm-free conducting tissues.

Ecologists have found that as the amount of CO_2 in air has increased in modern times, growth rates of plants and phytoplankton have increased as well. Recent data also suggest that during the last ice age, when atmospheric CO_2 levels were lower, global primary productivity was also reduced. Thus, CO_2 availability does appear to play a role in primary productivity, although any fertilization effect of carbon dioxide operates within the constraints of other limiting nutrients.

In chapter 2, we saw how the carbon cycle is intertwined with those of iron, sulfur, nitrogen, and other elements involved in metabolism. The Redfield ratio and its variations underscore another intimate dance that connects carbon to other cycles—in this case, nutrients. We can discuss different elements individually, but in nature, organisms and environments are molded by the aggregate and interactive physical and biological processes that cycle carbon, nutrients, and key players in energy metabolism in space and time.

CHAPTER 4

Completing Life's Alphabet

S and Trace Elements

The English alphabet has 26 letters, used with varying frequency in communication. Some—for example, *A*, *T*, *S*, and *N*—occur in many, perhaps even most, words. Others, like *Q* and *Z*, are much less common, but we need them, too. Absent *Q* there could be no Queequeg in *Moby-Dick*, and without *Z*, what would we call those striped horse cousins that roam African plains? The elemental alphabet of your body also runs to about two dozen letters (the exact number varies from one source to another), and like those of the English alphabet, some are abundant and others rare (figure 4.1). Again, however, we need them all to function.

We've already introduced the elemental cycles of C, N, and P, which together account for about 22 percent of a cell's weight and collectively govern ecosystem function on land and in the sea. Actually, oxygen (O) and hydrogen (H) are even more abundant in our bodies, as water (H_2O) makes up 60 percent or so of each cell's mass. O and H also occur in diverse biomolecules, including proteins and nucleic acids, and so contribute about 30 and 10 percent, respectively, to the dry weight of cells—less than carbon but far more than any other element. Half a dozen other elements

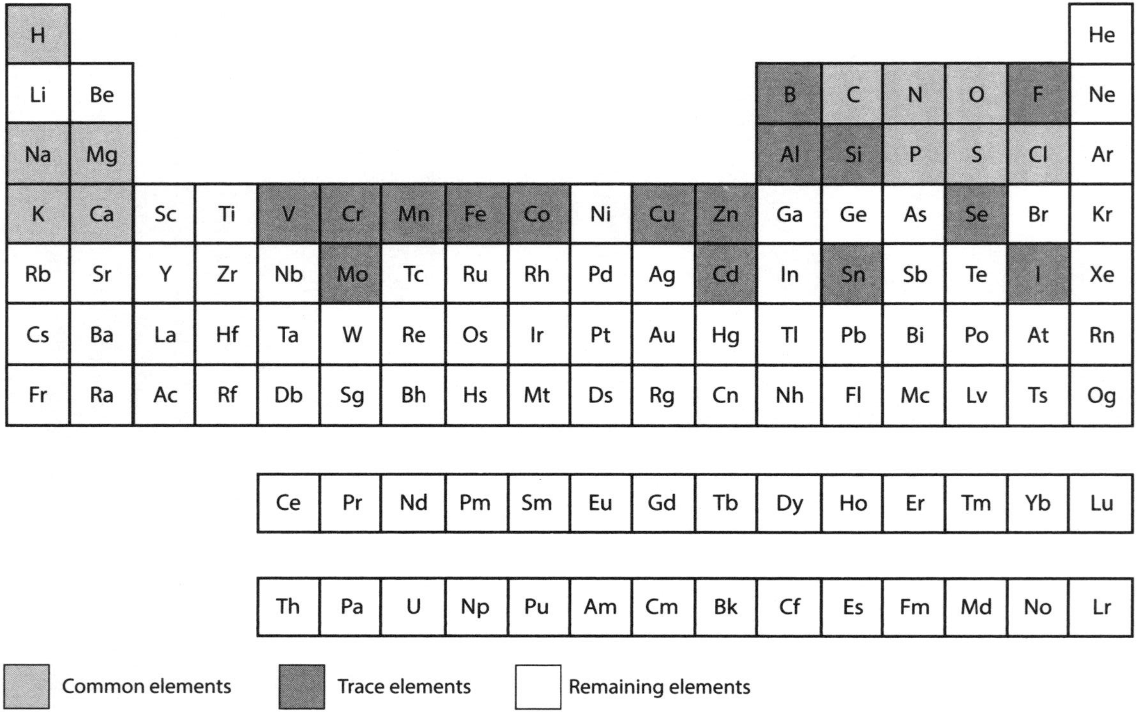

H																	He
Li	Be											B	C	N	O	F	Ne
Na	Mg											Al	Si	P	S	Cl	Ar
K	Ca	Sc	Ti	V	Cr	Mn	Fe	Co	Ni	Cu	Zn	Ga	Ge	As	Se	Br	Kr
Rb	Sr	Y	Zr	Nb	Mo	Tc	Ru	Rh	Pd	Ag	Cd	In	Sn	Sb	Te	I	Xe
Cs	Ba	La	Hf	Ta	W	Re	Os	Ir	Pt	Au	Hg	Tl	Pb	Bi	Po	At	Rn
Fr	Ra	Ac	Rf	Db	Sg	Bh	Hs	Mt	Ds	Rg	Cn	Nh	Fl	Mc	Lv	Ts	Og
				Ce	Pr	Nd	Pm	Sm	Eu	Gd	Tb	Dy	Ho	Er	Tm	Yb	Lu
				Th	Pa	U	Np	Pu	Am	Cm	Bk	Cf	Es	Fm	Md	No	Lr

Common elements   Trace elements   Remaining elements

FIGURE 4.1. The periodic table of the elements, showing those elements found as major, minor, and trace elements in cells.

are moderately abundant, each contributing 0.1 percent or more to the cell's weight, while the remaining 16 or so occur only in trace amounts (table 2.2). As we've already seen, rarity doesn't make these trace elements unimportant. Whether it be iron, cobalt, zinc, or iodine, all are essential. And all of them cycle between Earth and life.

In this chapter, we'll introduce the cycles of sulfur and the trace elements, all central to life and environments on Earth (and likely, wherever life may be found), saving oxygen and hydrogen for chapter 5.

Sulfur is commonly associated with the underworld, probably because sulfurous gases curl upward from volcanoes and hydrothermal fissures, while sulfur-bearing minerals, most prominently the garish yellow of native sulfur, precipitate around their edges and along the margins of hot springs. In reality, sulfur (S) isn't that common in the solid Earth, making up only about 0.7 percent of its mass, mostly sequestered deep in our planet's core. That said, sulfur shares several key features with carbon, and these underpin its biological importance. First, sulfur, like carbon, can be a solid, a gas, or dissolved ions at Earth surface temperatures. And second, again like carbon, sulfur is sensitive to its redox environment, resulting in compounds as oxidized as sulfate (SO_4^{2-}) and as reduced as hydrogen sulfide (H_2S), all useful in microbial metabolism.

If you collect a beaker of seawater and let it evaporate, salts will precipitate on the beaker's floor and walls. About 86 percent will be sodium chloride (NaCl), or table salt. Another 9 percent will be gypsum ($CaSO_4 \cdot 2H_2O$), along with a smidgen of Epsom salt ($MgSO_4 \cdot 7H_2O$). So seawater contains a fair amount of sulfur, as sulfate ions (SO_4^{2-}) in solution. Volcanic gases contain sulfur dioxide (SO_2) and hydrogen sulfide (H_2S). Both are injected into the atmosphere when volcanoes erupt, but they soon react with oxygen and rain out as sulfate ions, so air doesn't contain much sulfur. You, on the other hand, are relatively enriched in S compounds; sulfur makes of about 0.2 percent of your cells by weight.

Sulfur plays an essential role in your body and the bodies of all other organisms, occurring, among other molecules, in several amino acids; vitamin B7; and ferrodoxins, molecules that shuttle electrons back and forth within cells. Sulfur also contributes to nitrogenase, the key enzyme in microbial nitrogen fixation. As ever in the Earth system, sulfur enters the biota when sulfate ions are taken up from soil or water by primary producers. (Hydrogen sulfide is toxic to most organisms, although at exceedingly low concentration, it can serve as a signaling molecule within cells.) Grazers get the sulfur they need by consuming primary producers, and predators, in turn, eat the grazers, moving S along with C, N, and P through food webs. Dead organisms decompose, returning bioavailable sulfur to the environment. But, as was true for nitrogen, sulfur's biological importance extends well beyond its incorporation into biomolecules (figure 4.2).

We've already seen that where oxygen gas is absent, other oxidants are used in respiration. In the modern ocean, the most important of these oxidants is sulfate. Diverse sulfate-reducing bacteria use sulfate to respire organic matter within anoxic marine environments, accounting

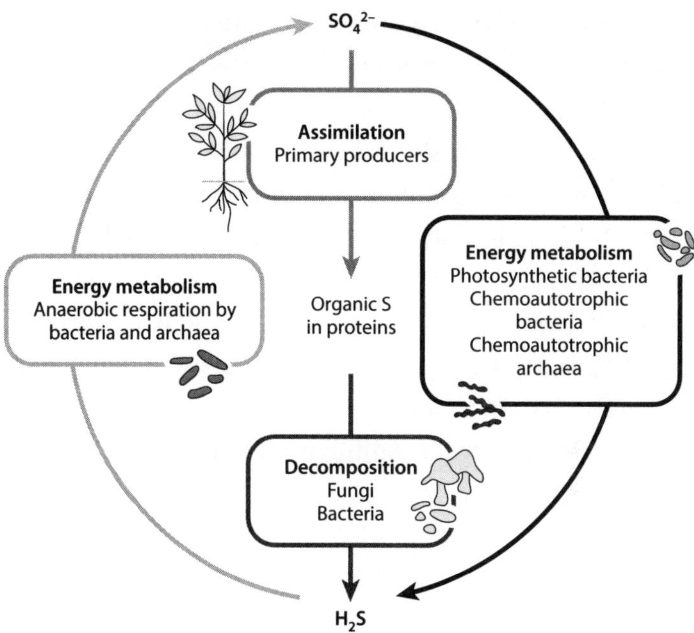

FIGURE 4.2. The biological sulfur cycle. Sulfur plays a key role in biological molecules, and because of its redox properties, it is also important in microbial metabolisms, especially in environments where oxygen gas is limited or absent.

for as much as 50 percent of the remineralization of organic matter delivered to seafloor sediments. Bacterial sulfate reduction produces hydrogen sulfide as a by-product, and this, in turn, provides electrons needed for carbon fixation in photosynthetic and chemosynthetic bacteria, regenerating sulfate in the process. Geologic processes, in turn, remove sulfur from the biosphere. For example, as seawater evaporates, sulfate ions are incorporated into gypsum (or its water-free cousin, anhydrite), sequestering it in sediments. In similar fashion, H_2S generated by sulfate-reducing bacteria can react with iron in solution, precipitating another mineral—pyrite (FeS_2), commonly known as fool's gold—within oxygen-free sediments. These minerals may eventually be thrust upward as mountains rise, and both the dissolution of gypsum and the oxidation of pyrite will restore sulfate ions to seawater. Alternatively, sulfur-bearing minerals may be delivered back into the mantle, eventually to return to the Earth's surface as volcanic gases. In short, physical and biological processes cycle sulfur through the Earth system, closely tied to Earth's biological carbon cycle.

It is worth noting that sulfur enters the conversation between Earth and life in one more distinctive way. As mentioned earlier, there isn't a lot of sulfur in the atmosphere, but there is a little, and it has important environmental consequences. During the 1970s, James Lovelock, father of the Gaia Hypothesis (chapter 1), showed that marine algae produce a compound called dimethylsulfoniopropionate (mercifully abbreviated DMSP). Within the water column, bacterial and photochemical reactions transform DMSP to dimethyl sulfide, or DMS, a gas that can escape to the atmosphere. Sulfate ions and sulfuric acid produced by the oxidation of DMS provide nuclei for the condensation of water vapor to form clouds. Extending this insight in an influential 1987 paper, Lovelock and colleagues argued that because clouds reflect incoming radiation, DMS production plays a significant role in regulating climate, a view now widely accepted (more about climate regulation in chapter 14).

The remaining elements in your body can be divided into two groups. First, there are electrolytes, simple ions whose biological importance stems largely from their positive or negative charges. And then there are metals, found mostly as cofactors in proteins. (Cofactors are ions or molecules whose incorporation into proteins is essential for enzyme function.) Electrolytes occur both within cells and in the bodily fluids that surround cells and tissues. The principal electrolytes are calcium (Ca^{2+}), potassium (K^+), sodium (Na^+), chlorine (Cl^-), and magnesium (Mg^{2+}), all of which play a major role in maintaining homeostasis, the internal conditions that facilitate cellular function. Electrolytes pumped into and out of cells regulate the osmotic balance of cells (not too much water or too little), as well as intracellular pH, important for many biochemical reactions. Beyond this, each electrolyte has a number of unique biological roles. Calcium (Ca), for example, is critical for signaling within and between cells and also functions as a cofactor in proteins that form blood clots. And in many organisms, Ca plays another key role, in the formation of mineralized skeletons, such as the calcium carbonate of clam shells and corals, as well as the calcium phosphate that hardens your bones (see chapter 12).

Sodium and potassium ions are particularly important for maintaining cellular homeostasis. Sodium levels are also key to healthy blood pressure; that's why your doctor urges you to moderate your intake of table salt. Chloride, negatively charged ions of chlorine, also contributes to cell homeostasis and helps to modulate oxygen and carbon dioxide levels in blood. Finally, there is magnesium. In your body, Mg works as a cofactor in several hundred different enzymatic reactions, and it contributes to the normal function of

ATP, your cells' energy currency. In plants, algae, and photosynthetic bacteria, magnesium plays another, altogether different role, lying at the functional heart of the chlorophyll molecules that capture light energy for photosynthesis. All of these electrolytes are readily available in the food and fluids you take in, and either too much or too little of any of them can result in disease. The electrolytes in soil and water, in turn, originate via the chemical weathering of rocks.

The term *metal* doesn't sound very biological. We're more likely to think of it in terms of foundries and factories, gold coins, or copper wiring. But like electrolytes, metals—quite a few of them—are critical, if minor, constituents of your cells.

Iron (Fe) made a brief entrance earlier, both in terms of energy metabolism and as a partner for sulfur in pyrite. At first glimpse, iron seems to be the most physical of elements, used by humans from the Iron Age to the iron horse, but not necessarily part of us. That, however, is incorrect. Iron is a minor component of your body; its abundance is measured in parts per million (about 60, by most estimates), rather than by percentage, but you can't function without it. Most of the iron in your body is found in hemoglobin, the molecule in red blood cells that enables you to transport oxygen to your brain, your toes, and all points in between. You also have a related molecule called myoglobin, another iron-binding protein that governs oxygen supply to your muscles. Additionally, iron is found in a number of other proteins: Some are bound to S and essential for electron transfer in respiration, while others are involved in the synthesis of the key structural protein collagen, neurotransmitters, and more. Indeed, iron is so important that you keep an internal store of it in molecules called ferritin—enough, it is said, to last you about three years if you are male and six months if female. Low iron in your blood leads to the most common form of anemia.

Like other elements, the iron you need comes from the food you eat, iron that entered the food web via uptake from the environment by primary producers. And therein, as Shakespeare almost said, lies the rub. As we discussed in chapter 2, iron is strongly sensitive to the presence or absence of oxygen gas. When O_2 is absent, iron in its reduced form (Fe^{2+}) can exist as dissolved ions in wet soil or water. Such conditions have only a limited distribution today, but as we'll see in subsequent chapters, most water in the oceans was anoxic for most of our planet's history. Today, however, the oceans are, for the most part, well oxygenated, and under these conditions, iron is rapidly oxidized to Fe^{3+} (ferric iron) and precipitates as iron hydroxides, or rust. As a result, the amount of iron in seawater is exceedingly low, so low that, as already noted, it can limit primary production. To cope, cyanobacteria and algae in the oceans have evolved a number of sophisticated strategies for scavenging iron from their environment. Most of the tiny amount of iron in seawater occurs as Fe^{3+} bound to organic molecules, and photosynthetic organisms are thought to import these complexes into their cells, dissociating Fe^{3+} from its organic partner at the cell surface and transporting it into the cell interior, or reducing the iron at the cell surface and importing it as Fe^{2+} (reduced iron). Many bacteria synthesize and secrete molecules called siderophores, which can bond securely with iron in a form that the cell can import into its interior. Algae and cyanobacteria commonly also take advantage of bacterial siderophores in their immediate environment, and some actually form symbiotic relationships with siderophore-secreting bacteria, giving them privileged access to some of the sequestered iron. And some land plants generate molecules that function much like the siderophores in bacteria.

Stunning images from the deep seafloor show that iron continues to be injected into the ocean at places where hydrothermal fluids vent upward into seawater. (The mid-Atlantic rift system, where oceanic crust is formed, widening the Atlantic Ocean, serves as a prime example; see chapter 7.) Although existing in its reduced form deep within the Earth, the iron that emanates from these vents is not dissolved Fe^{2+}. Ascending iron reacts quickly with oxygen gas to form oxides or with hydrogen sulfide to form pyrite and so enters the deep ocean in particulate form. Complexed with organic matter produced by microbes within the vent environment, this iron can travel through the deep sea for hundreds if not thousands of kilometers. Eventually, much of this particulate iron will rain out onto the seafloor, but some will make it to the surface, where it can be utilized by primary producers, much as they take in iron particles introduced by wind and attached to siderophores.

Because it can exist in either an oxidized or reduced form, iron can provide or accept electrons for energy metabolism in bacteria. It has long been known that some bacteria are able to use Fe^{3+} for respiration, reducing it to Fe^{2+}.

And in 1993, German microbiologist Fritz Widdel and his colleagues described, for the first time, photosynthetic bacteria capable of deriving electrons from ferrous iron, generating Fe^{3+} in the process. Like sulfur, then, iron metabolism can be coupled to the biological carbon cycle. That coupling plays only a subordinate role in modern ecosystems, but on the early Earth, when reduced iron was relatively abundant in seawater, it likely played a major role in cycling carbon through the biosphere—Earth's first Iron Age.

Iron may have a privileged place among the metals found in cells, but it is hardly alone. Vanadium, chromium, manganese, cobalt, copper, zinc, selenium, molybdenum, cadmium, selenium, and tin are all toxic at high concentrations but critical to the function of many proteins at the tiny abundances found within cells. Copper, for example, is a cofactor in various enzymes involved in energy metabolism, perhaps especially in cytochrome c oxidase, which effects the final step in respiration. Copper-bearing superoxide dismutase, in turn, mitigates damage from oxidative stress within cells. Copper also functions in cell signaling and is needed for iron absorption. Zinc, in turn, has myriad functions within cells; in our own bodies, for example, it is required for smell, taste, and the proper functioning of the immune system. Much of the zinc in our bodies—and, indeed, in cells of all types—lies in an enzyme called carbonic anhydrase, which catalyzes the reversible reactions between carbon dioxide and water, mediating the chemical conditions within cells. Unlike iron, these other metals do not play an important role in energy metabolism.

Silicon (Si), a major component of Earth's crust and mantle, is one more element that is attracting increased attention from biologists. Silica (SiO_2) has long been known to form mineralized skeletons in a number of organisms (see chapter 12). Siliceous skeletons are not widespread among animals, but some sponges form spicules of silica, and some choanoflagellates, the closest protistan relatives of animals, form basket-like skeletons of SiO_2. Land plants commonly have tiny tablets of silica in the walls of surface cells, thought to strengthen the cells at minimal metabolic cost. And among protists, silica bimineralization is actually widespread, especially among radiolarians, some of the most diverse protists in the oceans, and diatoms, tiny algae responsible for an estimated 20 percent of all photosynthesis on Earth. By themselves, diatoms keep dissolved silica levels in surface seawater at exceedingly low values.

All of these organisms have transporter proteins that bring Si into cells and tissues, and interestingly, recent genomic research shows that Si transporters are widespread among eukaryotic cells, even in organisms for which Si use is not known. Indeed, your body contains enzymes that transport Si into cells. The precise biological uses of Si in humans remain poorly understood, but increasing evidence associates Si with health in bones, joints, connective tissues, and more. Like other elements, Si cycles between Earth and life, entering the biosphere through weathering of rocks and being taken up by primary producers and organisms that fashion silica skeletons. Si leaves the biological silica cycle as largely skeletons deposited on the floor of lakes and the sea and in clay minerals that form or accumulate as mud. Eventually, these may be uplifted into mountains and reintroduced to the biological cycle by weathering and erosion, or they may be subducted and brought back to the surface in lavas.

My father-in-law, a lovely man much missed, was fond of (slightly mis-) quoting the Psalmist. “We are fearfully and wonderfully made” regularly escaped his lips. And so we are. Our cells, and the cells of all other organisms, contain two dozen or more elements, each playing essential and specific roles in our biology. Every atom in your body was once part of the physical Earth and will be again. It is this continuous conversation between Earth and organisms that sustains life on our planet and has done so for some four billion years.

CHAPTER 5

Water and Oxygen

Water is a fundamental feature of the habitable Earth, so much so that NASA long characterized the search for extraterrestrial life as "follow the water." About 97.5 percent of the H_2O at or near the Earth's surface is seawater, and most of the remainder consists of glacial ice, permanent snowpack, or groundwater. Lakes, rivers, and moist soils—the water of daily experience on land—make up only about 1 percent of Earth's fresh water (that's 1 percent of the 2.5 percent of the Earth's surface water not found in the sea). There's a bit of water vapor in the atmosphere and, in relative terms, just a smidge in organisms. Scientists debate how much water resides more deeply in Earth's crust and mantle, but reasonable estimates suggest that there is at least an ocean's worth of water in our planet's interior, perhaps more. Like carbon and nitrogen in the deep Earth, this water exchanges with surface reservoirs only on geologic timescales, delivered to the depths by subduction and returned to the surface in volcanic fumes.

Like grand mountain chains, Earth's great water bodies—the Pacific Ocean, Lake Superior, and the mighty Amazon—exude an air of permanence. That constancy is relative, of course. Even on a timescale of human lives, the water in these features is peripatetic, to say the least. To begin, seawater is in continual motion, circulating both horizontally and vertically through ocean basins. Cold seawater that sinks below the surface near Antarctica will cross the equator as a northward-trending, deep-sea current many decades later. Warm currents of the Gulf Stream travel northward along the east coast of North America before turning east to warm western Europe; palm trees in Scottish gardens owe their existence to this circulation.

Surface currents, strongly influenced by wind and constrained by the shape and connectivity of ocean basins, have been known to mariners for centuries. For example, early European explorers of the New World relied heavily on westward-flowing currents driven by trade winds that blow consistently from the Canary Islands toward the Caribbean. In contrast, the three-dimensional structure of deep-ocean circulation has become clear only in more recent times, facilitated, it turns out, by aboveground nuclear tests. Bomb tests carried out by several nations after World War II generated a large amount of radioactive carbon (the rare carbon isotope ^{14}C), some of which rained onto the sea surface. Since then, ocean scientists have traced the movement of these ^{14}C-spiked waters as they traverse ocean basins—as I write these lines, bomb carbon has made it to the deepest parts of the marine realm.

Pioneering research by the great oceanographer Wallace Broecker showed in detail how water masses move vertically as well as horizontally through ocean basins, forming what he called the great ocean conveyer belt. This whole-ocean transport is more formally termed thermohaline circulation, reflecting the importance of temperature (*thermo-*) and salinity (*haline*) in shaping observed flows. (Salinity and temperature play a major role because they are principal determinants of seawater density; denser water masses sink, while less dense waters rise toward the surface.) On the whole, ocean currents transport heat from low latitudes toward the poles, exerting a strong influence on global climate. In turn, upwelling, the return of deep seawater to the surface, supplies critical nutrients for primary production.

But water doesn't just slosh through ocean basins. It leaves the sea entirely, as evaporation draws water vapor upward to form clouds and, eventually, precipitation. Much like seawater, air circulates both vertically and horizontally. Near the equator, air masses warm, lowering their density and so causing them to rise. This warm air carries a great deal of water vapor, but as it rises, it begins to cool, and the water vapor condenses to form raindrops—that's why we see rain forests near the equator. Once equatorial air masses reach altitudes of 10–15 km (about 6–9 miles), they no longer continue to rise but instead spread toward the poles. At about 30° latitude, these now-cooled air masses descend again toward the Earth's surface, warming as they go. The descending air absorbs water vapor, rather than releasing it as rain, which goes a long way toward explaining the latitudinal distribution of our planet's great deserts. Some of the descending air will be directed back toward the equator, while the rest continues to move poleward, repeating the cycle of rise and fall as it goes. Actually, air currents don't move straight north or south. Because of the Earth's rotation, air in the Northern Hemisphere is deflected to the right as it moves; in the Southern Hemisphere, it deflects to the left. Called the Coriolis effect, this explains why trade winds blow westward at low latitudes while mid-latitude winds, appropriately called westerlies, move from west to east. Mountains play a role, too, causing air currents to rise and so to cool and drop precipitation as they run up against mountain ranges, and then warm and take up water vapor as they descend on the other side. The parts of Oregon, Washington State, and southern British Columbia that lie west of the Cascades are a lot wetter than lands east of this range.

Much of the atmospheric water vapor sourced from the oceans rains or snows back to the sea, but some carries moisture to continents and islands, connecting the huge reservoir of the sea to the fresh waters that sustain life on land. The hydrological cycle—the cycling of H_2O among water bodies, land, life, and air—is well studied on the continents for the simple reason that it is critical to understanding biological diversity and promoting human welfare. Water evaporates from lakes, rivers, and wet soils just as it does from the oceans, returning to the land as precipitation. Some of the precipitation will feed rivers and lakes; some will end up in glaciers, permafrost, or the oceans; some will seep into the Earth, sustaining subsurface aquifers; some will evaporate again; and some will be taken up by organisms (to be returned to the environment via excretion or death). Both rivers and rain deliver water from terrestrial reservoirs to the oceans.

Plants play a particularly important role in the hydrological cycle on land. Water absorbed by roots from wet soils moves upward through plant axes via the plant's vascular system, the wood in trees. Plant cells need water, but most of the water plants take up moves straight up through the stem and escapes as water vapor from leaves, never entering the plant's cytoplasm. Up and out—why should this be? For plants, the land surface is a desiccating environment; plants need to keep from drying out. Crucial to meeting this need is a surface coating of waxy material called cuticle, which retards water loss. Cuticle keeps water vapor from escaping from tissues, but it also forms an effective barrier against the atmospheric carbon dioxide that plants need for photosynthesis. The evolutionary solution was stomata, pores in the plant's cuticle and surface tissues that allow CO_2 to diffuse into photosynthetic tissues. But if CO_2 can enter the plant, water vapor can get out. The large volume of water that moves upward through the plant, thus, largely serves to balance the loss of water vapor during carbon uptake for photosynthesis. This system is sophisticated; environmental signals direct stomata to open and close in response to the varying needs to gain carbon or conserve water. Called transpiration, the amount of water that passes upward through plants and into the atmosphere as water vapor can be prodigious; in a single day, an acre of corn can issue 10,000–15,000 L (3,000–4,000 gallons) of water to the surrounding air. In the end, currents of water and wind, precipitation and evaporation, and biological processes cycle water through the biosphere, sustaining Earth as a habitable planet.

Like water, oxygen gas (O_2) is fundamental to life as we know it, at least the conspicuous life represented by plants and animals. Without O_2, you quickly asphyxiate; your body stops working as it runs out of the oxygen needed for respiration. At the same time, ozone (O_3) formed in the upper atmosphere by interactions between solar radiation and oxygen gas protects us from DNA-scrambling ultraviolet rays. Like other elements we've discussed, Earth's oxygen lies mostly in rocks of the crust and mantle. In fact, oxygen is the most abundant element in these depths, making up nearly half the mass of crust and mantle and accounting for more than 99 percent of all of our planet's total oxygen. The oxygen in Earth's interior is not breathable gas, and although we've already noted that the crust and mantle account for a large proportion of Earth's water, H_2O also contributes only a small fraction of the O found at depth. In fact, nearly all of the oxygen in Earth's interior is tightly bound to silicon (Si), forming the wide array of minerals that make up rocks of the crust and mantle (quartz, for example, is SiO_2). This oxygen exchanges little with the atmosphere and oceans.

Earth's atmosphere, hydrosphere, and biosphere together contain less than 0.05 percent of our planet's total mass of oxygen. Most of this resides in water, so the relative abundances of oxygen in oceans, glaciers, and fresh water are much like that of H_2O itself. (Because oxygen atoms have a much higher atomic weight than hydrogen, water is nearly 90 percent O by weight.) So, in one sense, the oceans, lakes, rivers, and glaciers are full of oxygen. Most of these water bodies also contain oxygen gas in solution—without it, fish couldn't swim in lakes or the sea—as well as oxygen atoms in various dissolved gases and ions, many of which we've already encountered, especially CO_2, HCO_3^-, CO_3^{2-}, SO_4^{2-},

H_2S, and PO_4^{3-}. As discussed earlier, O is a major constituent of cells, largely as water, but it also occurs in a diversity of organic molecules. And oxygen makes up nearly 21 percent of the air we breathe, mostly as O_2, but also including a bit of O in carbon dioxide, water vapor, and trace gases.

The cycles of water and oxygen gas are tightly linked within the carbon cycle. When we first introduced the biological carbon cycle in figure 2.1, we focused on carbon itself: how photosynthesis converts environmental CO_2 to organic matter and how respiration oxidizes organic carbon back to CO_2. At the same time, however, we noted that the same biological reactions cycle oxygen between O_2 and H_2O. Indeed, photosynthesis is the principal source of oxygen gas to the environment. Small amounts of O_2 can be generated by abiological processes. For example, water molecules in the upper atmosphere can be dissociated into hydrogen and oxygen gases by ultraviolet radiation from the sun. The hydrogen molecules are light enough to escape from the grip of Earth's gravity, but oxygen gas stays put, resulting in a net increase of O_2 in the atmosphere. And recently, several interactions between water and minerals have been shown to generate both oxygen gas and hydrogen peroxide (H_2O_2). Such processes may account for the small amounts of oxygen gas on Mars. (Mars's fabled red color is basically rust.) They cannot, however, account for the large volumes of oxygen gas in the atmosphere and dissolved in both marine and fresh waters. For that, we need photosynthesis.

For years, biologists assumed that the O_2 generated via photosynthesis came from CO_2, but careful experiments in the 1940s showed otherwise. Most O is ^{16}O, with eight protons and eight neutrons. A small amount is ^{18}O, with two additional neutrons, and an even smaller proportion is ^{17}O, with one extra neutron. By growing plants in experimental chambers in which either CO_2 or H_2O was enriched in ^{18}O and then measuring the isotopic composition of the oxygen gas released during photosynthesis, biologists were able to demonstrate that light energy breaks down water, freeing hydrogen for the reduction of carbon dioxide and releasing oxygen gas to the environment. Aerobic respiration, then, provides the other side of the biological carbon and O_2/H_2O cycles, reacting oxygen gas with organic molecules to generate energy and releasing water as a by-product. So, indeed, oxygen gas in the air you breathe was produced by photosynthetic organisms, and the O in atmospheric O_2 was once found in water and will be again.

Looking again at other elemental cycles, we see once more that oxygen plays an important role. As microorganisms respire organic carbon in O_2-poor environments, sulfate (SO_4^{2-}) is reduced to hydrogen sulfide, its oxygen atom transferred to H_2O; anaerobic photosynthesis using H_2S as a source of electrons moves it back again. Similar metabolic two-steps in the nitrogen and iron cycles also shuttle O between oxidized compounds and water. In short, in both oxic and anoxic environments, oxygen continually moves between oxides and water, cycling carbon while supplying energy for life.

If you try to quantify the amount of O_2 produced each year by primary producers and the amount removed by aerobic respiration, you'll find that the two numbers are similar—similar but not equal. The reason, as noted in chapter 2, is that some organic matter evades respiration and accumulates in sediments. Photosynthetic O_2 production thus exceeds O_2 loss via respiration, raising the possibility that atmospheric oxygen levels might increase through time. There is, however, another set of processes to consider, physical processes at play in the Earth system. Volcanoes and hydrothermal systems release reduced gases into the environment, and these react with O_2 in air and water, providing a geologic sink for oxygen gas. The chemical weathering of minerals provides another sink, as reduced ions such as Fe^{2+} react with ambient oxygen gas. And, as noted earlier, in the long run O_2 can get another crack at that buried organic matter, as long after its burial the organic materials in sedimentary rocks can be exposed to oxidation via weathering and erosion. Less intuitively, the burial in sediments of pyrite (FeS_2), formed when iron reacts with H_2S generated by bacterial reduction of sulfate (SO_4^{2-}), also contributes to the O_2 supply of the atmosphere and oceans.

Stepping back, we see that on the modern Earth, the production and consumption of O_2 more or less balance each other out. More broadly, we see that the various elemental cycles discussed in previous chapters are all linked by C, O, and H_2O in the Earth system. In many ways, we can't understand any one of the component cycles without placing it in the framework of the others.

CHAPTER 6

Trimming the Tree of Life

My name appears in the credits for precisely one movie, Terrence Malick's luminous *The Tree of Life*. In *The Tree of Life*, Terry suggests, obliquely to be sure, that the physical and moral actions that pattern our lives reflect a deep evolutionary framework from which we only partially escape by means of grace. At the time he made this film, Terry and I had been discussing another project, on the history of Earth and life. So if you follow *The Tree of Life* credits to the bitter end, just before the logos of Dolby and the Motion Picture Association of America cue the house lights, you'll find a gracious note of thanks for advice on life's deep history.

When Terry titled his film, he tapped into an ancient and powerful archetype. At least since the time of early Mesopotamian civilizations, the Tree of Life has conveyed a sacred sense of universality in cultures around the world. To biologists, the concept has a different meaning, although it still carries an unmistakable sense of universality, underscoring the unity of the living world despite its immense diversity. What the biological Tree of Life conveys is a sense of the evolutionary relationships among all species on Earth, an indispensable road map for life's diversification across some four billion years. Life may have arisen any number of times in our planet's infancy, but all life that we know about—bacteria, amoebas, redwoods, and us—traces back to a single ancestral population.

Charles Darwin understood the tree's metaphorical significance for evolution. In a notebook entry from 1837, more than two decades before publication of *On the Origin of Species*, he sketched a little stick-figure tree, accompanied by the brief but penetrating words "I think" (figure 6.1). *On the Origin of Species* itself contains but a single figure, one that depicts the tree-like relationships among species expected to result from natural selection, as well as the sense that many ancient species no longer exist. The accompanying discussion is memorable:

> *The affinities of all the beings of the same class have sometimes been represented by a great tree. I believe this simile largely speaks the truth. The green and budding twigs may represent existing species; and those produced during each former year may represent the long succession of extinct species. At each period of growth all the growing twigs have tried to branch out on all sides, and to overtop and kill the surrounding twigs and branches, in the same manner as species and groups of species have tried to overmaster other species in the great battle for life. The limbs divided into great branches, and these into lesser and lesser branches, were themselves once, when the tree was small, budding twigs; and this connexion of the former and present buds by ramifying branches may well represent the classification of all extinct and living species in groups subordinate to groups. Of the many twigs which flourished when the tree was a mere bush,*

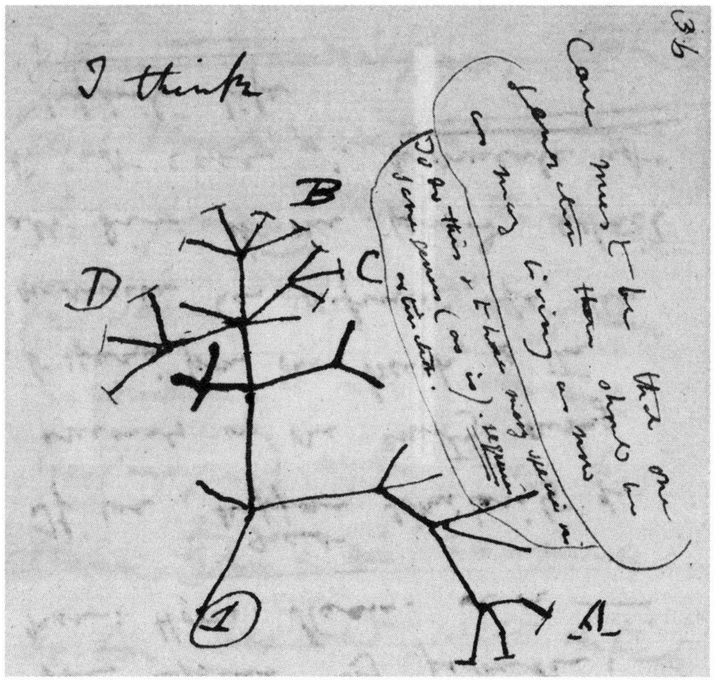

FIGURE 6.1. Twenty-two years before Charles Darwin published *On the Origin of Species*, he sketched the evolutionary relationships among organisms as tree-like in form. "I think," he noted. Indeed, he did.

> *only two or three, now grown into great branches, yet survive and bear all the other branches; so with the species which lived during long-past geological periods, very few now have living and modified descendants. . . . As buds give rise by growth to fresh buds, and these, if vigorous, branch out and overtop on all sides many a feebler branch, so by generation I believe it has been with the great Tree of Life, which fills with its dead and broken branches the crust of the earth, and covers the surface with its ever branching and beautiful ramifications. (Darwin, 1859)*

In the preceding chapters, myriad organisms with diverse metabolisms were introduced. The biological Tree of Life allows us to place this diversity in an evolutionary framework, to trim the tree, if you will, showing both how distinct metabolic pathways are distributed among Earth's rich biota and how they arose through time.

Biologists conventionally depict evolutionary relationships among different groups of organisms as a tree-like branching diagram, or phylogeny. For example, figure 6.2 shows a phylogeny of primates, based on comparisons of the genomes found within mitochondria. At the right end, we see that humans and chimpanzees fall on shallow branches that diverge from a single point called a node. This means that humans and chimps are postulated to be more closely related to each other than either is to any of the other primates shown in the figure. The node from which they diverge represents their last common ancestor. In turn, the branch ending in humans and chimps diverges from a node shared with gorillas—the common ancestor of all three species (plus bonobos, the first cousins of chimps), but no others. And so it goes as we move downward through the tree, ending up at the base with a trunk that features the last common ancestor of all living primates as its first branch point. The phylogeny thus presents a picture of the evolutionary relationships among the various primates shown in the tree, with closeness of relatedness indicated by recency of common ancestry. The tree also indicates the relative timing of evolutionary events, with branches near the distal tips understood as more recent than those lower down on the trunk. Using fossils of known age to calibrate

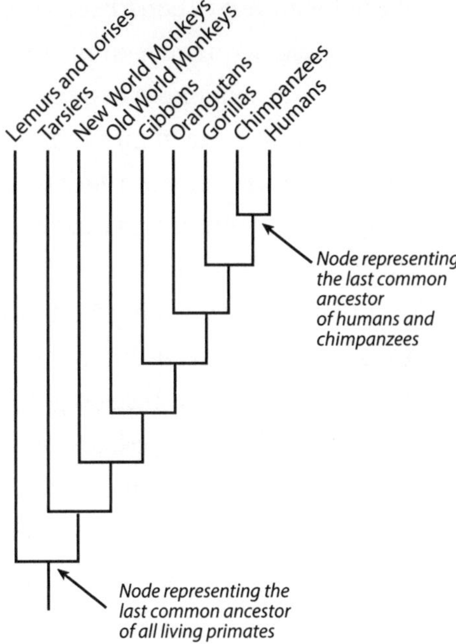

FIGURE 6.2. A phylogeny of primates, showing a hypothesis of evolutionary relationships among living members of the family to which we belong. Nodes are the points from which distinct groups diverge. The nested branching conveys a sense of both time and closeness of evolutionary relationship.

the tree, we can estimate when in our planet's history these evolutionary divergences took place.

That's all well and good, but how do we actually determine which taxa are most closely related to each other? In simplest terms, the more similar two species are, the more closely they are related. Chimps and humans are considered to be more similar to one another than either is to gorillas because they share features not found in gorillas or any other species. Readers will be able to make a list of morphological features that unite humans and chimps, but nowadays, most phylogenies are constructed using molecular data: the amino acid sequences of proteins or, more commonly, the nucleotide sequences of genes, even whole genomes. It is important to bear in mind that two species can display common features for three distinct reasons. First, as just noted, shared features may have arisen in the common ancestor of the two groups and will not occur in other groups. That's what we're looking for. But species can also share features that arose earlier in the tree. Humans and chimps both have hair, but then so do gorillas and, indeed, all mammals. Thus, while the presence of hair tells us that humans and chimps are both mammals, it does not help us to sort out evolutionary relationships within primates. And sometimes, traits evolve independently in two groups that are not closely related. The classic example is bats and birds; both have wings, but many other features make it clear that their wings evolved in very different ways from distinct ancestors. Such features are termed convergent, and they also don't help us to construct accurate trees.

In practice, analysis of the traits used to construct phylogenies requires computers that can provide estimates of the probability that any given hypothesis of branching provides the best fit to available data. Uncertainties remain, but our understanding of the Tree of Life is far better today than it was 50 years ago.

Figure 6.3 shows what is called a universal phylogeny—that is, a phylogeny that encompasses all of life's diversity from bacteria to humans. For some preliminary orientation, find the branch on which you reside—the animals. All animals,

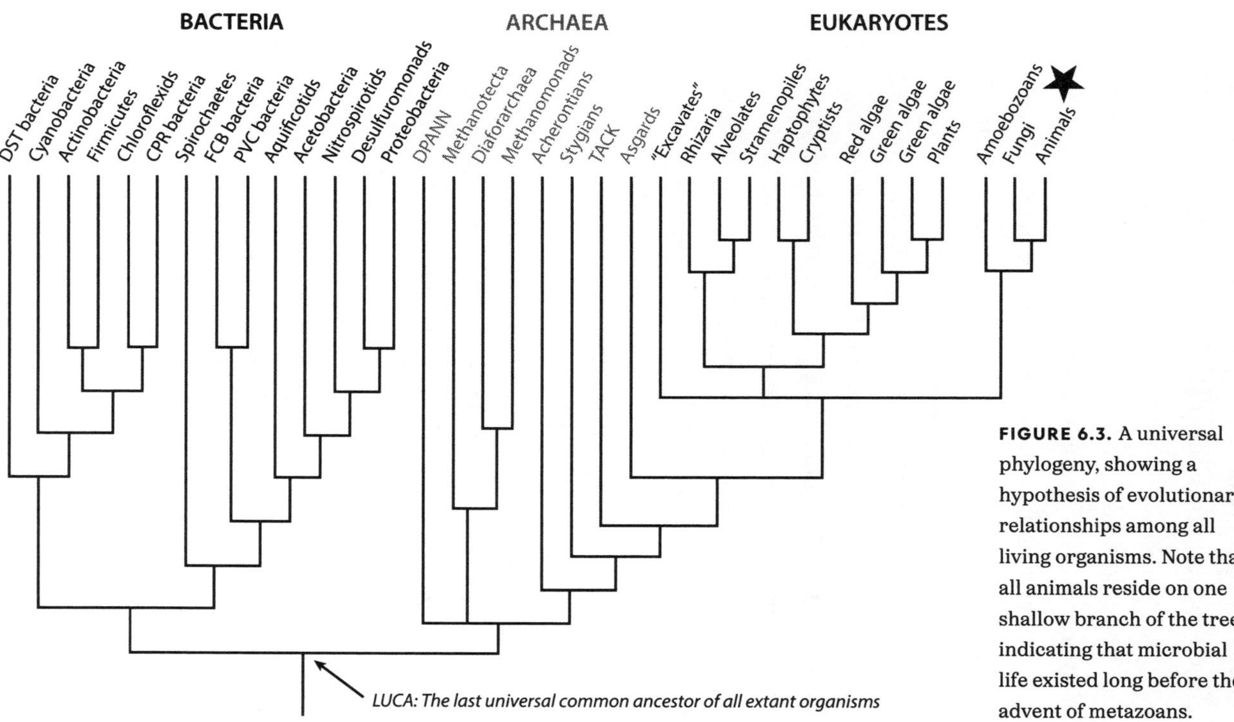

FIGURE 6.3. A universal phylogeny, showing a hypothesis of evolutionary relationships among all living organisms. Note that all animals reside on one shallow branch of the tree, indicating that microbial life existed long before the advent of metazoans.

from tiny placozoans to savvy octopuses and hyperdiverse insects, occupy this one distal branch of the tree. The first lesson, then, is that animals make up only a small part of Earth's biological diversity. Many of the branches, and all of the deep ones, are microbial.

Not surprisingly, the discovery of Earth's microbial world followed on the heels of a technological advance: the microscope. Although neither actually invented the microscope, two seventeenth-century individuals did much to improve it and, importantly, trained this new tool on nature. In his seminal work *Micrographia*, published in 1665, the British naturalist Robert Hooke presented the first known image of a microorganism, a tiny fungus. About a decade later, Antonie van Leeuwenhoek, a Dutch draper who fashioned high-quality lenses to help in the close work his trade required, discovered what he called "animalcules" in some kitchen liquid that had been sitting around for several weeks. Animalcules, as he soon figured out, are pretty much everywhere; most are now recognized as bacteria and microscopic eukaryotes. Hooke himself confirmed Van Leeuwenhoek's discoveries.

Only in the nineteenth century would the importance of this microbial world to both nature and medicine begin to emerge. A number of scientists proposed that bacteria caused disease, and the great Louis Pasteur not only confirmed this so-called germ theory but developed vaccines for several bacteria-borne illnesses. On the same timescale, the pioneering microbiologists introduced in chapter 1 began to understand how diverse microbial metabolisms help to shape ecology and environments. But what were these tiny creatures—degenerate animals, perhaps, or something completely different from our familiar macroscopic biota? Informed by improved microscopes and new insights into biochemistry, the French biologist Édouard Chatton proposed in 1937 that all organisms can be divided into two groups based on the organization of their cells. Prokaryotes, the bacteria known to Chatton and his contemporaries, consisted of tiny cells without a membrane-bounded nucleus or the complex organelles found in better-known organisms, such as plants and animals. Chatton recognized that the microbiota found in soils, the sea, and pretty much everywhere else was full of prokaryotes, but most environments also contained cells with nuclei, which he classified as eukaryotes, some unicellular and others made up of many nucleated cells. Bacteria, then, were not tiny offshoots of plants and animals; they are distinct from familiar life at the most fundamental level of cell organization.

Still, the concept of bacteria remained murky, with little idea of which bacteria were related to which. In an influential 1962 paper, R. Y. Stanier and C. B. van Niel called this out as "an abiding scandal" that stood in the way of developing a deeper understanding of microbial life. They argued, but could not prove, that all bacteria are descended from a common ancestor—that is, they form a distinct branch on the Tree of Life. The test came over the next decade, through the research of Carl Woese, one of the few individuals I've known who could legitimately claim to have changed the face of biology. A few intrepid scientists had begun to use the sequences of amino acids in proteins to determine genetic relatedness among organisms, but for Woese, proteins evolved too rapidly and were insufficiently universal to sort out bacterial phylogeny. Instead, he focused on RNA, specifically the RNA molecules found in ribosomes, the sites within cells where nucleic acids are translated into proteins. Woese knew that all organisms contain ribosomal RNA and also that it changes only slowly through time—just what he needed to study bacteria. Today, such work is routine, but when Woese began it was extraordinarily tedious, requiring both skill and determination. Slowly, species by species, Woese built the first bacterial tree, in the process laying the groundwork for a Tree of Life that encompasses all life on Earth.

As related in a fascinating article by Virginia Morell, a pivotal moment came in 1976, when Ralph Wolfe, a colleague of Woese's at the University of Illinois, suggested that he include methanogens, a group of microorganisms that produces methane, in his analysis. As Morell tells it:

> *When Woese studied their sequences, the methanogens did not register as bacteria. "They were completely missing the oligonucleotide sequences that I had come to recognize as characteristic of bacteria," he explains. Thinking the sample had somehow been contaminated, he ran a fresh one. "And that's when Carl came down the hall, shaking his head," says Wolfe. "He told me, 'Wolfe, these things aren't even bacteria.' And I said, 'Now, calm down, Carl; come out of orbit. Of course, they're bacteria; they look like bacteria.'" But, as Woese now knew, morphology in bacteria meant nothing. Only their molecules told the story. And the molecules proclaimed that the methanogens*

were not like any other prokaryote or eukaryote—they were something unto themselves, a third branch of life. (Morell, 1997)

The following year, Woese and then postdoctoral fellow George Fox published an astonishing paper. Not only was it possible to construct a universal Tree of Life, they wrote, but the tree had three main branches: bacteria, eukaryotes, and a previously unappreciated third limb, now known as the archaea (home of, among other groups, those nonconforming methanogens).

One more remarkable innovation fueled the gathering revolution in microbiology. Norman Pace, a microbiologist gifted with both genius and curiosity, reasoned that if scientists could extract and sequence nucleic acids from microbes found in laboratory cultures, they could do the same for microorganisms in environmental samples. Once again, the results were astonishing. When Norm began his work, a few thousand species of bacteria were known, each one laboriously isolated and maintained in pure culture (no other species present). In hindsight, it is evident that most microbes in nature have strong interdependencies with other species and so cannot readily be cultured. And so, as Pace and others began to probe soil and water samples, it became clear that most natural populations had never been cultured; the true diversity of bacteria and archaea ran to the millions, not the thousands. In time, biologists learned to extract complete genomes from environmental samples, providing a new sense of not only who is out there but how they make their livings. This revealed diversity of the microbial world fundamentally changed our understanding of the interactions between physical and biological processes in the Earth system.

As yet, there is no consensus on branching order within the bacteria and archaea. In part, this is because key events happened long ago, with subsequent gains, losses, and alteration of genes complicating phylogenetic analysis. And then there is the issue of horizontal gene transfer. In high school biology, you may have learned that genes pass vertically from one generation to the next; you got your genes from your parents and will pass them on to your children. Bacteria and archaea, however, commonly exchange genes with distantly related microbes. This can occur by the direct transfer of genes from one microbe to another in physical contact; microbes can also take up genes released from dead cells; and viruses commonly transfer genes from one cell to another. Studies of bacteria and archaea show that an estimated 90 percent or more of all their gene families have passed horizontally between distantly related species at least once. (Eukaryotes engage in horizontal gene transfer as well, but much less commonly.) Because of this, species that are not closely related can contain genes that are.

Despite these problems, biologists have made increasingly sophisticated efforts to reconstruct evolutionary relationships among microbes. The skeletal tree shown in figure 6.3 has several key features that are likely to prove correct however the details turn out. First, the deepest branch point among living organisms is the split between bacteria and archaea. This basal node is not the earliest organism; much simpler, but now extinct, microbes must have lined the path from the origin of life to the last common ancestor of all living organisms (see chapter 8). Comparison of features found in bacteria and archaea tells us that the last common ancestor of all living organisms was prokaryotic; the intracellular complexity of eukaryotes would come later. The last common ancestor was also unicellular; only later would cells adhere to one another to form sheets, filaments, and, eventually, complex multicellular forms with distinct tissues and organs. And early branches on both the bacterial and archaeal limbs indicate that their common ancestor was anaerobic, consistent with geologic evidence that oxygen gas was rare or absent on the early Earth (see chapter 9). On the other hand, bacteria and archaea have distinct types of lipids in their membranes, and the details of how nucleic acids give rise to proteins differ as well. In these cases, we really can't tell what was present in the last common ancestor.

What about metabolism? The universal common ancestor was probably chemoautotrophic, using chemical reactions to provide the energy needed to fix CO_2 into organic molecules; photosynthesis, as well, would come later. That said, two examples illustrate the challenges and promise of trimming the Tree of Life with metabolic pathways. As introduced earlier, microbial sulfate reduction plays a key role in both the carbon and sulfur cycles. Genes that code for the various reactions involved in microbial sulfate reduction are widely distributed on the tree. Thirteen distinct groups of bacteria and archaea, found on widely separated branches, are known to harbor these genes. That might suggest that microbial sulfate reduction evolved early in the history of life and then was lost in most groups. Or perhaps the ability to respire organic matter using sulfate ions was acquired independently in many groups via horizontal gene transfer. If we make a tree using only genes involved in sulfate reduction, we find that the tree is rooted within the archaea and only later spreads to bacteria. This being the case, it appears that microbial sulfate reduction evolved within archaeal microbes and then spread into bacteria by repeated episodes of horizontal gene transfer.

Photosynthesis, unequivocally the most important metabolism ever to evolve on Earth, also tells of both antiquity and horizontal gene transfer. Like microbial sulfate reduction, photosynthesis is widely distributed among bacteria—as noted in a previous chapter, at least eight distinct groups are known to use light energy to fix carbon and/or gain energy. While not a primordial metabolism, photosynthesis evolved early, allowing bacteria to spread across the globe. Most photosynthetic bacteria do not use water as a source of the electrons needed for carbon fixation; nor do they release oxygen gas as a by-product. Horizontal gene transfer appears to have spread the constituent parts of photosynthetic systems from one bacterial group to another. By acquiring a combination of molecular features that enabled them to use water as an electron source, one group of bacteria, the cyanobacteria, hit the photosynthetic jackpot, eventually to change the face of the Earth. In time, the oxygen gas they generated began to accumulate in the atmosphere and surface ocean (see chapters 9–11). Other oxidants such as sulfate also increased in abundance, helping to explain the relatively late expansion of diversity among sulfate-reducing bacteria.

The archaeal limb is also coming into focus, as new types of archaea are discovered each year. The earliest archaea were not only anaerobic but quite possibly thermophilic (that is, living at elevated temperatures). As in the bacteria, archaea diversified to include a wide variety of metabolisms, including the previously mentioned methanogenesis, a uniquely archaeal metabolism that must have evolved early within the group. In contrast, no photosynthetic archaea are known to exist, perhaps related to the distinct composition of their membranes. A particularly important discovery was made in 2015, when an entirely new group of archaea called the Asgards was discovered in samples from the vicinity of a hydrothermal ridge in the Arctic Ocean. Members of this group harbor genes similar to those that govern a number of key features of cell biology in eukaryotes, and, increasingly, it looks like Asgard archaea or their close relatives played a major role in the origin of eukaryotic cells.

As noted earlier, eukaryotes form a group with limited metabolic diversity but an astonishing ability to change cell shape and to form complex multicellular bodies. Eukaryotes differ markedly from bacteria and archaea in many aspects of cell biology, but just as clearly, they are descendants of the universal common ancestor. When Woese and Fox originally argued for a Tree of Life with three branches, they left branching order unspecified. Within a few years, however, evidence emerged in support of the hypothesis that eukaryotes were more closely related to archaea than they are to bacteria. Eukaryotes and archaea, it seemed, diverged from a common ancestor, and that ancestor diverged from bacteria in the basal branching of the tree—a universal tree with two great dichotomies, giving rise to the three major domains of life.

But there was more to come. As both the amount of genomic data available and the ways of analyzing it mushroomed, evidence mounted that eukaryotes emerged from *within* the archaea. And with the discovery and characterization of Asgard archaea, this rendering of the tree (figure 6.3) has become canonical. Again, however, the story is more complicated. The last common ancestor of living eukaryotes had a mitochondrion, the membrane-bounded organelle responsible for aerobic respiration in eukaryotic cells. Already with the discovery of mitochondria in the late nineteenth century, it was proposed that these organelles originated as once free-living bacteria that had been captured and reduced through time to metabolic slavery. The idea was radical, and in the absence of means to test it, it didn't gain favor among biologists. By the mid-twentieth century, however, emerging cell and molecular biology provided just the required tools. Thus, when Lynn Margulis, then a young biologist at Boston University, rearticulated the hypothesis in 1967, the stage was set for one more revolution in biological understanding.

Lynn proposed that the mitochondria, chloroplasts (the organelles responsible for photosynthesis in eukaryotes), and flagella of eukaryotes all originated as bacteria that were captured by relatively simple proto-eukaryotic hosts. Her "endosymbiotic hypothesis" (endosymbionts being symbiotic microbes that live *within* a distinct host cell) could now be tested because both mitochondria and chloroplasts have their own DNA and so can be placed on trees built from genetic data. When the tests were run, it became clear that molecular-sequence comparisons place mitochondria within a group of bacteria called the proteobacteria and chloroplasts within the cyanobacteria. That is, both organelles are descended from once free-living bacteria. (The part of the hypothesis dealing with flagella didn't gain similar support.) To the extent, then, that eukaryotes form a third limb on the Tree of Life, it is not simply because of sequential divergences, but, rather, the reflection of a unique and consequential merger of cells from the archaeal and bacterial branches. Indeed, eukaryote genomes contain genes acquired from many different groups of bacteria, and some biologists argue that a series of endosymbioses gave rise to the eukaryotic cell.

The last common ancestor of living eukaryotes was an aerobic cell capable of acquiring nutrition via phagocytosis—the engulfment of food particles from the environment. Thus, early eukaryotes included predators, a novel way to gather food that would transform ecology. Later, the common ancestor of red and green algae captured a cyanobacterium, bringing photosynthesis to eukaryotes. Photosynthesis would eventually spread across the eukaryotic limb by further endosymbiotic events involving photosynthetic symbionts that were, themselves, eukaryotes. A relatively small number of eukaryotes live in oxygen-free environments, mostly using modified mitochondria to gain energy, but in general, eukaryotic organisms, whether large or small, participate in the biological carbon cycle through aerobic respiration and oxygenic photosynthesis.

Having trimmed the Tree of Life with metabolic pathways, we can pull back and view the tree as a historical record of life's conversation with the physical Earth. Clearly, the biological processes involved in the dialogue between Earth and life have changed through time. At the dawn of life, a limited diversity and abundance of anaerobic cells populated Earth, relying on chemical reactions to fix carbon dioxide into the organic molecules needed to sustain our planet's nascent biosphere. It is widely thought that microbial nitrogen fixation also evolved early, although abiotic fixation of nitrogen by lightning and other physical processes may have been sufficient to supply early cells with the required inventory of bioavailable N. In due course, some organisms developed the ability to use solar energy to fix carbon, sourcing needed electrons from a variety of environmental molecules, such as hydrogen gas, hydrogen sulfide, and reduced iron in solution. Life expanded beyond the local environments where chemoautotrophs could thrive, and as one group of photosynthetic bacteria evolved the capacity to use water as a source of electrons, new forms of life spread across our planet's surface.

As oxygen gas accumulated in the atmosphere and surface ocean, aerobic respiration emerged as a globally significant part of the carbon cycle. At the same time, oxidized ions such as sulfate and nitrate increased, expanding the role of sulfate reduction and denitrification in other key elemental cycles. And with the evolution of eukaryotic cells, the stage was set for complex multicellularity and, in time, sentience. As the tree makes clear, our familiar world of plants and animals is an evolutionary latecomer, preceded by a long antecedent history of microbial life. Even today, there are nearly 30 tons of bacteria for every ton of animals on Earth. And for much of our planet's history, life was completely microbial.

CHAPTER 7

Living on a Dynamic Planet

As an undergraduate (gasp) half a century ago, I enrolled in a course on structural geology, the study of folds, faults, and other features that record the deformation of Earth's crust. We learned to analyze features that reflect compression, as seen, for example, in the folded beds exposed in the Front Range of Colorado. We also studied fault systems that reflect tensional forces—think of the Great Rift Valley of eastern Africa. And we came to appreciate faults whose two sides move in opposite directions relative to one another; Californians know all too well the San Andreas Fault and the havoc its movements spawn. Most of all, we learned about plate tectonics, the theory that explains all these observations and much more. What we didn't learn—and which disappoints me to this day—was that at the time of our course, plate tectonics was an emerging concept, a totally new way of understanding the Earth that was both exciting and controversial. New ideas and the debate they engender provide wonderful educational opportunities, not to be wasted. My undergraduate experience in that course informed my own teaching throughout my long career.

Had I taken my course a decade earlier, we would have learned to describe faults and folds in much the same way; that was worked out decades earlier. Their explanations, however, would have been entirely different, hopeful hypotheses sustained more by necessity than by evidence. What changed during the 1960s was scientific attitudes toward an old idea: that continents move through time, with mountains and rift valleys recording episodes of continental collision and breakup. In 1915, a German meteorologist named Alfred Wegener laid out the idea of continental drift in a retrospectively famous book entitled *The Origins of Continents and Oceans*. Today, Wegener is celebrated as the prophet of a new tectonics, but at the time most Earth scientists simply considered him mad. Wegener's hypothesis of drifting continents seemingly required that these great masses of granite plow across or through the basaltic crust that underlies Earth's seafloor. "No way," said the critics, and the idea was largely relegated to a dark scientific closet.

Largely, but not entirely. In South Africa, a gifted geologist named Alex du Toit saw many features in the field that supported Wegener's claim. Du Toit knew, for example, that South African rocks from the Permian Period (290–252 million years ago) contain fossil evidence for a distinctive land biota, with cool temperate forests dominated by a broad-leaved tree called *Glossopteris* that shaded diverse, now extinct, mammal-like reptiles. Some 2,000 miles across the Atlantic Ocean, Permian rocks in South America recorded much the same ecosystem, as did comparably aged rocks in peninsular India, Australia, and, eventually, as exploration accelerated, Antarctica. Du Toit also recognized similar patterns of physical geology on either side of the South Atlantic and found it easiest to explain all these Southern Hemisphere observations if he could retrospectively close up the ocean, as Wegener had proposed. Despite widespread admiration for Du Toit's talents, his writings didn't garner much international support for continental drift.

Our modern understanding of the physical Earth began to emerge in the 1950s, with the careful mapping of the seafloor beneath the Atlantic Ocean. Bruce Heezen, an oceanographer at Columbia's Lamont-Doherty Earth Observatory, repeatedly traversed the Atlantic, collecting detailed data on the ocean floor. At Lamont, his colleague Marie Tharp used these data and others to draw the first accurate map of the Atlantic seafloor. Remarkably, Tharp's map showed that a mountain belt runs down the length of the Atlantic Ocean, submerged, save for Iceland, beneath the sea.

Princeton geologist Harry Hess soon recognized that this distinctive feature, the Mid-Atlantic Ridge, could resolve the geophysical objections to continental drift. In a far-reaching 1962 paper, Hess argued that the ridge system was a place where magma rising from within the mantle formed new oceanic crust, which slowly but inexorably pushed North and South America apart from Europe and Africa. Continents do move, said Hess, but not because they plow through oceanic crust. Rather, the continents are embedded in rigid plates and move as those plates grow along oceanic ridge systems. Earth isn't getting larger, so plate destruction has to occur as well, and this was recognized along the margins of the Pacific Ocean, where slabs of oceanic crust sink beneath other plates, melting as they go. This explains why volcanoes—the product of all that melting—line much of the Pacific's margins.

Within half a dozen years, numerous geophysical observations supported and refined Hess's hypothesis, and plate tectonics was born. A limited number of rigid plates grow from oceanic ridges and are consumed in subduction zones. Mountains form where oceanic crust subducts beneath a continent, as in the Andes of South America, or when two continents collide—the Alps, for example, or the Himalaya. The African rift valley can be recognized as a place where tension has begun to split the great continent in two. And the San Andreas Fault is simply a place where two plates grind past each other. All of these crustal dynamics are rooted in the underlying mantle, which, as British geologist Arthur Holmes first proposed a century ago, convects over long timescales. Hot mantle rising toward the surface unpins the formation of new crust at spreading centers, while subduction zones coincide with places where cooler mantle sinks back toward our planet's core.

Plate tectonics, then, provides a first-order explanation for Earth's physical features. And, as geologists learn early on that "the present is the key to the past," plate theory has been used to interpret our planet's deep history. Plate tectonics provides a pretty good accounting for the rock record of the Phanerozoic Eon (the past 539 million years). Africa collided with Europe to give us the Alps, while India plowed into southern Asia, raising the Himalaya. Rifting initiated the Atlantic Ocean more than 200 million years ago, and since then it has grown to its current size. The Appalachian Mountains record an earlier collision between Africa and North America, and the Urals formed as Europe and Asia crashed together. Plate tectonics explains the distribution of *Glossopteris* fossils across the Southern Hemisphere, as well as both paleontological and magnetic evidence that some rocks in British Columbia and southern Alaska formed far from North America, later to be sutured to the continent.

Cycles of continental collision and breakup can be traced backward into much earlier times, but as we descend deeper into the past, cracks start to develop in the assumption of tectonic invariance. It's not that we lose all evidence of plate tectonics (although that has been suggested), but the details start to change. Subtle but illuminating evidence comes from a seemingly esoteric source: the strontium isotopes of carbonate rocks deposited on the seafloor. Stay with me here. Five isotopes of strontium are known from nature, most abundantly ^{87}Sr and ^{86}Sr. Strontium enters the oceans in fluids expelled at mid-ocean ridges and as ions weathered and eroded from continents. In general, ridge strontium has a relatively low ratio of ^{87}Sr to ^{86}Sr, whereas strontium in continental crust has a significantly higher ratio. When limestones form, they incorporate strontium ions from seawater, providing a record of the relative inputs of continents and ridges to the ocean's chemical makeup through time. Today, the ratio of ^{87}Sr to ^{86}Sr in seawater is high, in no small part because of the weathering and erosion of ^{87}Sr-rich rocks exposed in the Himalaya. That ratio is also high in limestones deposited near the end of the Proterozoic Eon, about 600 million years ago, thought by many to reflect the uplift of ancient Himalaya-scale mountains (see chapter 11). In carbonate rocks deposited over the billion years preceding this event, however, ^{87}Sr/^{86}Sr values are low; landscapes were subdued, and so continents simply weren't supplying weathered materials to the oceans at high rates. Chris Spencer, at Queen's University in Ontario, has used this and other lines of evidence to argue that during what is commonly called Earth's middle age (roughly 1,800–600 million years ago), the crust was thinner and hotter than it is today, perhaps reflecting heat from the underlying mantle trapped beneath long-lived supercontinents. Mountains formed during this interval, but they didn't make grand peaks like those we see today. Earth's middle age was a protracted interval marked by low oxygen, the absence of continental glaciers, and low rates of evolutionary change. To explain such things, scientists increasingly look to the distinctive tectonics of the time.

Some sort of plate tectonics is thought to have operated throughout the Proterozoic Eon (2,500–539 million years ago), and some geochemical evidence suggests that high mountains formed early in the eon, before the onset of Earth's relatively quiet middle age. However, as we continue backward into the Archean and Hadean eons (over 2,500 million years ago), the preserved record gets smaller and less easily interpreted, spawning debates about when plate tectonics began and how it worked early on. Indeed, we have no terrestrial rocks older than four billion years, meaning that we have to reconstruct our planet's earliest history from meteorites—the materials from which Earth formed—along with truly ancient minerals preserved as grains in younger sandstones. Plus some pretty good physical theory.

Evidence from meteorites tells us that Earth began to accrete about 4.54 billion years ago, and that it heated up as it grew. Some heat reflects the transfer of kinetic energy as meteorites struck and stuck to our rocky protoplanet, and some resulted from the decay of then-abundant radioactive materials. In consequence, the infant Earth melted, and dense materials, especially iron, sank to form our planet's core, while minerals rich in silica (SiO_2) made up the overlying mantle. It is thought that the Earth's earliest surface consisted of molten materials much like the lavas that flow from volcanoes in present-day Hawaii—a magma ocean in geologists' evocative phrase. Within a few million years, this cooled to form a primordial crust, and so, in terms of tectonics, the game was on.

I mentioned ancient minerals found in sandstones of a younger era. The minerals in question are zircons, small crystals made of zirconium and silica ($ZnSiO_4$) that incorporate a bit of radioactive uranium as they grow. Radioactive isotopes of uranium decay to form lead, and, because we know the rate at which this occurs, we can calculate the mineral's age by measuring the amount of lead that has accumulated through time. Some of these minerals are as old as 4.4 billion years, and they tell four important stories about the early Earth.

Zircons form in Earth's interior when basalt-like rocks begin to melt. Not all of the minerals found in basalt melt at the same temperature, and the resulting partial melting gives rise to magmas that crystallize to form new types of rocks, more like granite than the parent basalt. Granite is the stuff of continents, and so, in at least a broad sense, ancient zircons record the earliest differentiation of granite-like crust. Microchemical details suggest that sediments carried downward into the Earth formed some of the source materials for these early zircons. Many scientists interpret this as evidence for subduction, and so some nascent form of plate tectonics. Other chemical details suggest the presence of fresh water at the time the source materials for the zircons formed. In aggregate, then, ancient zircons suggest (1) the early appearance of at least broadly continent-like rocks, (2) the early onset of subduction, (3) the antiquity of Earth's hydrological cycle, and (4) the early emergence of dry land. In short, the conditions envisioned for the origin of life (see chapter 8).

Not surprisingly, many of these interpretations have provoked vigorous debate. To begin, we must ask why early basalts melted. Some Earth scientists argue that this reflects subduction—and so some form of plate tectonics, perhaps facilitated by large impacts that cracked and submerged parts of the ancient crust. Others counter that long-lived volcanic fissures could have generated bed after bed of basalt, forming a stack that thickened to the point where rocks at the bottom of the pile began to melt. No plate tectonics required. Still others argue for something called sagduction. In this view, on the early Earth, thick piles of basalt essentially sank into less dense crustal rocks beneath them, again causing partial melting (as well as faulting and folding) without invoking plate tectonics. (Both overthickening of basalt flows and sagduction could incorporate sediments, weakening the argument that evidence for sediment input to early zircon formation requires subduction.)

Who is correct? Well, quite possibly everyone and no one, reminiscent of the famous line in John Godfrey Saxe's poem

"The Blind Men and the Elephant" that "each was partly in the right, / And all were in the wrong!" The problem is that scientists have sometimes envisioned an Earth system with two states: In one, there is plate tectonics, more or less like today. In the alternative, no plate-like processes were at work. Increasing research, however, strongly suggests that the aggregate of patterns and processes characteristic of modern plate tectonics evolved piece by piece over much of the Hadean and Archean eons. Lateral movements of Earth's crust have been documented in rocks as old as 3.25 billion years, but they may well have coexisted with overthickened or sagducting piles of basalts from which granites differentiated—distinct processes operating in different places at the same time. And while subduction probably originated early, on the young Earth, crust that subducted into hot mantle may have been brittle, breaking apart even as it began to sink, short-circuiting the process. What we can say is that during this long "early" interval of Earth history (actually, some 44 percent of the whole), plate tectonic processes emerged and gradually came to dominate global tectonics. Continents grew, eventually forming the kind of large, stable masses we know today.

And there was another physical consequence of note. If there is one thing that geologists agree on, it is that Earth's interior heated up in its youth and has been slowly cooling since that time. During the Hadean and Archean eons, therefore, our planet's mantle must have been hotter than it is today. As noted in chapter 5, something like an ocean's worth of water currently resides in the mantle, but early on, at the temperatures modeled by scientists, the mantle wouldn't have retained much water at all; most of Earth's water would have degassed to the surface. If you add another ocean's worth of water to the modern Earth surface, most land gets submerged. And it would have been worse on the early Earth because the hot mantle couldn't support the kind of high-standing continents we see today. Three and a half billion years ago, then, Earth was pretty much a water world, with only limited landmasses sticking out above a planetary ocean.

Through the long Archean Eon, continental crust—perhaps most of it—formed. Subduction increasingly returned degassed water back into the mantle, and as mantle temperatures cooled, more and more of that H_2O stayed there. The geological record indicates that through time, more and more continental crust emerged from the sea. By the end of the eon, we have evidence for large emergent continents on mobile plates. The physical Earth was growing more familiar.

But even as the basic tectonic processes that shape our planet rounded into form, the physical Earth remained dynamic. If we could view a movie of the Earth's surface taken through time from some orbital outpost, we would see a continual geographic dance as landmasses move across the surface. Through time, continents collided, aggregating to form expansive supercontinents four times in the past two billion years. Inevitably, these supercontinents fractured, and their constituent pieces separated as oceanic crust grew between them. Two hundred million years ago, most of Earth's continental crust was wrapped up in the supercontinent Pangaea; today, the continents are about as dispersed as can be. And, as continents sashayed from place to place, the shapes and orientation of ocean basins also changed, strongly influencing the transport of heat by seawater and, therefore, climate. Mountains rose and wore away, further impacting climates across the globe. As a result, Earth's climate, like its geography, has changed repeatedly, with episodic ice ages separated by essentially glacier-free intervals (see chapter 14). The physical Earth remains dynamic to this day.

The Tree of Life shows that biological voices in the conversation between Earth and life have diversified through time. The rocks beneath your feet tell us that Earth's physical voices have changed as well. Thus, the conversation itself has evolved through time, shaping the land and sea, the chemistry of the atmosphere and oceans, global climate, and life. Changes on one side of the conversation have repeatedly influenced the other side, with biology sometimes in the driver's seat, and physical processes taking the lead at other times. The product of this ever-changing conversation is our planet's history, and the history of the life it sustains.

CHAPTER 8

The Conversation Begins

We've characterized our planet in terms of a long-running conversation between two sets of voices, one physical, the other biological. It's been that way for some four billion years, but before that, in Earth's earliest days, there could have been only a single choir, fashioned by chemistry and physics. How did life arise on the young Earth, and how did it come to be an important set of planetary processes in its own right?

In 1859, Louis Pasteur put the final nails into the coffin of spontaneous generation, the long-held notion that life can spring spontaneously from nonliving materials and has done so regularly through time. Pasteur's experiments convinced him (and, soon, the scientific community in general) that organisms derive always and only from preexisting organisms. That solved one problem, but it opened up a bigger one: Where did the first organisms come from? This question was much on Darwin's mind, in part because a number of prominent biologists had argued that his theory of natural selection was incomplete without accounting for life's origins. In several letters and publications written during the 1860s, Darwin simply argued that the problem, while interesting, lay beyond the ken of science. However, in a famous 1871 letter to the botanist Joseph Hooker, Darwin gave rein to his imagination:

> *But if (and oh what a big if) we could conceive in some warm little pond with all sorts of ammonia and phosphoric salts, light, heat, electricity &c present, that a protein compound was chemically formed, ready to undergo still more complex changes. (Darwin, 1871)*

Perhaps, Darwin speculated, life really is the product of physical processes at play on the early Earth.

Questions of life's origins couldn't be answered in Darwin's day, but neither would they go away. Several essays in the 1920s framed origin questions within a mechanistic framework, pointing the way toward experimental studies to come. Both the Russian biochemist Alexander Oparin and British geneticist J.B.S. Haldane envisioned a primordial Earth enveloped in an oxygen-free atmosphere. Driven by energy from lightning, heat, and ultraviolet radiation, simple inorganic molecules combined to form organic matter, characterized by Haldane as a "hot dilute soup." Oparin further argued that as the concentration of organic molecules increased, they would spontaneously segregate into tiny droplets, which Oparin called coacervates. In his view, these emerged as the factories of life's origins, taking in energy and materials from their surroundings and slowly developing the capacity for metabolism. Haldane, in turn, emphasized the nascent ability of early structures to reproduce and so to evolve.

Things came to a head in the early 1950s, when Stanley Miller, then a graduate student in the laboratory of Nobel laureate Harold Urey at the University of Chicago, decided to put some of these ideas to the test. Miller combined water, methane (CH_4), ammonia (NH_3), and hydrogen gas (H_2) in a beaker and heated it to generate a gaseous mixture that approximated Urey's view of Earth's primitive atmosphere. Then, he ran a spark discharge through the mixture to simulate lightning as it arced through the early atmosphere. Within a week, the walls of the flask had turned a reddish brown—deposits of organic matter formed by chemical reactions in the beaker. Miller's analysis showed that the organic molecules he generated included amino acids, the building blocks of proteins. (Miller reported only three amino acids that he could identify with certainty, as well as two others that needed confirmation, but half a century later, reanalysis of his sealed beakers showed that some 20 amino acids had actually formed during the experiment.) Questions of life's origins were now open to experimentation.

In the decades since Miller's pioneering research, we've learned a great deal about how life may have taken root on the primitive Earth. That's not to say that the problem is solved—not by a long shot. But Stanley would be pleased at the progress in recent years.

More than 30 years ago, Steve Benner, an unusually creative scientist who works at the interface between chemistry and biology, discussed the emergence of life in terms of three key milestones. The road from Miller to Mahler didn't begin with his first milestone, and it didn't end with the last, but the events Steve highlighted form a useful road map for the establishment of life as a key component of the Earth system. To begin, there were the first entities that might reasonably be considered living—that is, structures that reproduced and evolved. Then came the "breakthrough" organism—the first to make use of proteins coded for by nucleic acids. And finally, at the other end of the tunnel, came the last common ancestor of all extant life. As we've already seen, this was a complex cell containing DNA, RNA, proteins, and more. Inferences about the last common ancestor rely primarily on phylogeny and comparative biology, but as we move backward in time toward the first replicating entities, experiments and theory gain prominence.

Organisms contain a host of complex molecules that collectively underpin cell function and replication, and nearly all of these molecules form from simpler building blocks. Proteins, for example, are built from amino acids; the amino acids—20 different kinds—become linked to form chains that fold into complex three-dimensional shapes, conferring the physical structure and catalytic ability that we associate with proteins. This, of course, is where Stanley Miller comes in, demonstrating, as he did, that amino acids can form spontaneously under plausible conditions on the early Earth. In the decades since Miller's first experiment, numerous studies have shown that amino acids can form in a variety of reasonable prebiotic environments. Moreover, amino acids have been detected in carbonaceous meteorites and even in interstellar space, making it clear that the relevant chemistry is not some rare piece of terrestrial good luck, but, rather, fundamental chemistry found widely in the cosmos.

Similarly, DNA and its molecular cousin, RNA, are complex molecules made up of simpler building blocks joined end to end. The structural units of DNA are nucleotides, each one formed from sugar, a phosphate ion, and a relatively simple organic molecule called a base. The bases, four of them, impart informational value to nucleotides, so that the linear sequence of nucleotides in a DNA (or an RNA) molecule encodes the information needed for the cell to function and reproduce. For many years, scientists have known how to synthesize sugars and bases from simple inorganic precursors, and more recently, success was achieved in generating nucleotides themselves, not by joining ribose sugar and a base with phosphate ions—that approach failed for decades—but by mixing together a number of simple organic molecules likely to have been present on the early Earth.

And the theme of large complex molecules built from simpler units continues to lipids, such as the fatty acids that join with an alcohol (glycerol) and a phosphate ion to form the membranes that segregate cell contents from their surroundings and, in many cases, maintain the structural integrity of a cell's interior. Like other molecules under consideration here, the several components of membranes can be synthesized under plausible prebiotic conditions and occur in carbonaceous meteorites.

The take-home point is that the complex molecules of life are built from simpler components, and decades of research shows clearly that those building blocks can form in a variety of ways under plausible prebiotic conditions. The game is not without rules: Oxygen gas must be absent, and the higher the mixture's ratio of H to C, the better. Energy needed to synthesize life's building blocks could have come from lightning, heat, high-energy solar radiation, or radioactive decay, all in ready supply when our planet was young. Joining these monomers (the technical name for the building blocks) is trickier, as formation of amino-acid or nucleic-acid polymers is difficult in the dilute oceanic soup envisaged by Haldane. Wetting and drying help tremendously, favoring synthesis along shorelines or in shallow lakes, and minerals such as clays or pyrite can provide templates for polymerization, as can certain simple organic molecules. Because of their polar nature (a positive charge on one end of the molecule and a negative one on the other), fatty acids form membrane-like bodies spontaneously, and these can encapsulate protein-like and nucleic acid–like molecules.

Getting from these prebiotic syntheses to real live organisms requires that the molecules within nascent cells be able to replicate themselves, store information, and catalyze the chemical reactions that sustain both reproduction and function. And if nascent cells were to grow and reproduce, they also needed to incorporate both materials and energy from their environment. That is, they needed some simple form of metabolism.

For many years, the question of how nucleic acids and proteins came to interact as they do today has been a major problem for origins researchers. In cells from yours to bacteria, the information needed to make proteins is encoded in DNA, and both RNA and proteins are required to convert this information into new proteins that function inside the cell. Because of this difficulty, discoveries by Thomas Cech and Sidney Altman in the early 1980s that some RNA molecules can function like enzymes made of protein rocketed through the origins-of-life community. (It also earned them the Nobel Prize in 1989.) If early RNA molecules could store information like DNA while catalyzing biochemical reactions as proteins do, the pathway to life was simplified, or so it seemed. Steve Benner's preferred ur-organism consisted largely of RNA molecules inside a membrane-like lipid container, with the RNA providing information storage, as well as the catalysts needed to read and translate that information into active molecules (also RNA). Then, as now, errors in the replication of the RNA-storage system generated heritable variations on which natural selection could act. The earliest populations could evolve.

Benner was aware that while an RNA-based ur-organism kicked the problem of protein incorporation down the road, it didn't eliminate it—thus, his next milestone: the "breakthrough organism." This was the first organism to synthesize proteins via the translation of information encoded in nucleic acids. Most cell functions remained the province of RNA molecules, but DNA was now emerging as a stable library of cellular information, while proteins began their protracted takeover of catalytic function. More complex metabolisms became possible, in many ways initiating the two-way conversation between Earth and life.

Many new ways of thinking have emerged since Benner published his provocative (in the best sense of the word) essay. Increasingly, for example, researchers are investigating scenarios in which hydrogen cyanide and related precursor molecules might simultaneously give rise to amino acids and nucleotides, raising the possibility that peptides and short nucleotide polymers interacted from the very beginning of prebiotic synthesis. Also, an old idea, recently resurrected, points toward cofactors as key players in moving life toward the breakthrough organism. Today, many proteins have associated structures called cofactors that play a key role in enzyme function. Many of these cofactors are structured much like nucleic acids—vitamin B12, for example, which is a cofactor required for DNA synthesis. Other proteins have metal ions at their functional hearts; iron sulfide cofactors play key roles in energy metabolism.

Conventionally, proteins are considered the heavy hitters in cellular catalysis, and cofactors merely helpers. But perhaps this gets things backward, at least from the perspective of biological history. Perhaps nucleic acid–like cofactors were principal catalysts in an early RNA-rich world, with proteins evolving to enhance their activity. And perhaps early redox reactions were mediated by minerals such as pyrite (an iron sulfide), which later became embedded within proteins, bringing their catalytic properties under intracellular control. Lots of perhapses here, but what we've learned so far offers a promising route forward.

And then there is the question of metabolism. Years ago, during a coffee break at an origins-of-life meeting, a friend noted that metabolism isn't necessary for life. "Yes," I replied, "but for many of us, metabolism is what makes life interesting." Steve Benner saw metabolism as part and parcel of the breakthrough organism, even though the nature of early metabolic pathways remained unspecified. Others, however, have not only specified the earliest form of carbon and energy metabolism but gone on to argue that it is central to the origin of life itself, with heritable information in the form of nucleic acids coming later.

Bill Martin, an innovative molecular biologist, has been a particularly ardent advocate for the deep origins of the reductive acetyl-coenzyme A, or Wood-Ljungdahl, pathway of biosynthesis. Some of the earliest branching microbes on both the archaeal and bacterial branches of the Tree of Life utilize this metabolic pathway to synthesize organic molecules from hydrogen and carbon dioxide. Interestingly, similar reactions occur abiotically in the depths of hydrothermal vent systems, catalyzed by minerals that include iron sulfides. Perhaps early metabolisms evolved as proto-organisms began to commandeer such physical processes, gradually, step by step, evolving enzymes to take over the process. In this view, the information storage provided by nucleic acids would have come later. I'll admit that I find it easier to accrete metabolic pathways onto organisms that already have an RNA-based genetic memory than the reverse, but the debate is by no means over. And, as noted above, origins researchers are increasingly taking seriously the idea that information and metabolism took shape together in the earliest stages of life's history. Stay tuned.

However we view the incorporation of metabolism into cells, complex metabolic pathways were present in Benner's third milestone organism, the last common ancestor of extant life. As we've noted, this was a complicated microbe, with DNA, RNA, and proteins, diverse enzymes and their cofactors, membranes, and the ability to use both energy and materials from its surroundings. Getting from the breakthrough organism to the last common ancestor of extant life is itself complicated, but it can be fairly viewed as the product of natural selection—evolution generating ever more complex protomicrobes and novel ways of responding to diverse environments.

When did all this happen? When did Earth emerge as a planet in which physical and biological processes interact to cycle carbon and other elements? Fortunately, life leaves a calling card in sedimentary rocks deposited through time. Bones document the immensity of dinosaurs that roamed the landscape 240–66 million years ago, and other fossils record animal life back to about 575 million years before the present. Of course, animals evolved into a biological world already rich in microbial diversity, and, perhaps surprisingly, many of these early microorganisms also imparted a decipherable record to sediments deposited during their lifetimes. Actual fossils of microorganisms can be traced back three billion years, with more controversial microstructures even older. Stromatolites, laminated sedimentary structures that reflect the interaction between microbial mat communities and physical sedimentation (figure 8.1) go back nearly 3.5 billion years, and probable isotopic signatures of metabolism occur in metamorphic rocks as old as 3.95 billion years. Even older, tiny flecks of carbon within a zircon crystal formed 4.1 billion years ago have a ratio of carbon's two stable isotopes, ^{13}C and ^{12}C, that is similar to that in photosynthetically derived carbon molecules found in younger rocks, although nonbiological alternatives are possible. The simple fact is that as we move backward in time, we run out of rocks before we run out

FIGURE 8.1. A stromatolite, its distinctive shape reflecting the interaction of physical processes with microbial communities, from late Archean (ca. 2.7-billion-year-old) carbonate rocks in Australia.

of evidence for life. What we learn from the geologic record, then, is that microbial ecosystems definitely existed 3.5 billion years ago, probably existed nearly four billion years ago, and possibly originated even earlier. Earth has been a biological planet for most of its long history.

If the "when" of origins takes us back four billion years or more (indicating that animals have been present for only the last 15 percent of life's history), what can we say about questions of where? Hints of "where" are scattered through the preceding paragraphs. Many scientists favor lakes or tidally influenced shorelines as likely locales for the origin of life, not least because they feature episodic wetting and drying and so allow ambient waters to develop high concentrations of prebiotically interesting molecules. Others, more interested in metabolism, look to deep-sea hydrothermal vents as places where early metabolic pathways could have originated. I'm fond of reminding my origins colleagues that in modern Iceland, the one place where the Mid-Atlantic Ridge rises above sea level, hydrothermal vents and lakes occur together. On the early Earth, which, as noted earlier, was probably something of a water world, much of the land that emerged from the oceans would have consisted of volcanoes with associated hydrothermal systems, so Iceland might be trying to tell us something. In any event, if, as seems reasonable, the first metabolisms were similar to the reductive acetyl-coenzyme A pathway found in extant methanogenic archaea and acetogenic bacteria, early life would have been a local phenomenon, tethered to environments where H_2, CO_2, and minerals to catalyze their interaction were available. Only with the evolution of photosynthesis would life expand across the planet.

Earth's primordial atmosphere may have resembled the mixture used in the Miller experiment, but collision with a Mars-sized body 30–60 million years after the planet began to accrete would have stripped away this early gaseous envelope—it also formed the moon! The succeeding atmosphere, formed as gases continued to emanate from the nascent mantle, is thought to have consisted largely of nitrogen gas, carbon dioxide, and water vapor—no oxygen gas, but limited concentrations of reducing gases like hydrogen. Meteorite impacts, however, could have injected reduced gases into the atmosphere, repeatedly creating a Miller-friendly atmosphere, at least transiently. Additionally, or alternatively, hydrothermal systems could have maintained local environments that facilitated prebiotic chemistry. Once life emerged, continuing "prebiotic" chemical reactions would simply have provided food for early microbes.

While many uncertainties remain, we've learned enough to be comfortable with the idea that life on Earth is a product of physical processes at play on our young planet. As early microbes diversified, their evolving metabolisms established the conversation between the physical and biological Earth that persists to this day.

CHAPTER 9

The Story of O, Part 1

Oxygen is critical to human survival. Deprived of O_2 for even a few minutes, you will die. Our bodies have some ability to acclimate to oxygen levels lower than those generally encountered by most people—think, for example, of the remarkable Sherpas who scale Mount Everest, where oxygen levels are only about a third those at sea level. Such tolerance, however, is limited. And it's not only humans; nearly all animals perish without O_2 in ready supply. Now place that thought alongside one considered in the previous chapter: When life began, there was no oxygen gas in Earth's atmosphere.

Only without oxygen gas could life have originated on our planet, but only with O_2 could animals and many other organisms evolve. The contrast is striking, and it leads inevitably to a key historical conclusion: Some time after life first appeared at the Earth's surface, oxygen gas must have permeated the atmosphere and oceans, creating environments capable of supporting novel forms of life, eventually including us. What was the source of this O_2? As discussed earlier, there are a number of nonbiological ways to generate oxygen gas. Mars, for example, is called the Red Planet because its surface is tinted by iron oxides, rust formed on a lifeless landscape. Similar processes may well have been at work on the early Earth, but none seems capable of generating enough O_2 to transform our planet. We know only one process capable of achieving this: photosynthesis by organisms that split water molecules to obtain the electrons needed to fix CO_2, releasing O_2 as a by-product. That is, those remarkable photosynthetic bacteria called cyanobacteria and their descendants, the chloroplasts in algae and land plants. Stepping back, then, we see that not only do many forms of life need oxygen in their environment, but on our planet, oxygenated environments need life. Few if any stories in Earth history better illustrate the interplay between physical and biological processes than the story of O.

The cliffs lining Dales Gorge glow fiery red as the sun sets over northwestern Australia (figure 9.1). A highlight of Karijini National Park, set deep within the Australian Outback, the gorge has long been a magnet for tourists in search of nature's beauty. Geologists admire the beauty, too, but they also appreciate Dales Gorge for another reason. It tells a story about the young Earth.

In part, the gorge's almost surreal color reflects oxidized iron minerals formed recently, as underlying rocks, exposed by erosion, react with water and oxygen gas. But brush away the surface dust and you'll still see red. The cliffs consist of iron and chert (SiO_2) in alternating layers. Geologists call the rocks banded iron formation: banded because of their regular layering, and iron formation because these are among our planet's richest sources of iron ore. The story of iron formations hinges on a simple observation: Sediments like those preserved in Dales Gorge

FIGURE 9.1. Iron formation in Dales Gorge, Australia. The 2.5-billion-year-old rocks consist of interlayered silica and iron minerals, deposited on an ancient seafloor beneath oceans far different from those at present.

cannot form in modern oceans. As introduced in chapter 4, iron is one of those elements whose solubility depends on the presence or absence of O_2 in ambient waters. Today, oxygen pervades the oceans from surface layers in direct contact with the atmosphere to bottom currents flowing over the deep seafloor. Under these conditions, iron is basically insoluble. Today, Fe^{2+} released into seawater rapidly oxidizes to form rust. Indeed, as noted earlier, in present-day oceans, iron is so scarce that it limits photosynthesis over wide swaths of the sea.

Some 2.5 billion years old, the iron formation at Dales Gorge documents a different ocean, one in which iron could move long distances through the sea before being deposited on the seafloor. And Dales Gorge is hardly unique. Banded iron formations occur globally, but for the most part only in rocks older than 2.4 billion years. (There were significant but relatively short-lived recurrences about 1.86 billion years ago and, again, some 717–635 million years ago.) The difference between pre- and post-2.4-billion-year-old sedimentary rocks? Oxygen.

Actually, several lines of evidence indicate that our planet was oxygen-poor in its youth. As observed in surface sediments at Dales Gorge, on the modern Earth, iron released from exposed continental rocks by weathering and erosion readily oxidizes to form rust. Rivers and wind deposit the oxidized iron minerals along with sand and mud, imparting a reddish hue to accumulating sediments—red beds, in geological parlance (figure 9.2). The jagged rocks that frame Red Rocks Park and Amphitheatre in Colorado provide a memorable example. Interestingly, red beds are little known from sedimentary rock successions older than 2.4 billion years, but they are common features of younger successions—just the opposite of iron formations. The explanation: oxygen again, this time largely in the atmosphere.

Other features reinforce this story. Take pyrite (fool's gold, FeS_2), for example. Pyrite is highly sensitive to oxygen. Exposure to O_2 at even low concentrations results in its rapid oxidation to form sulfate (SO_4^{2-}), commonly deposited as gypsum. In ancient sandstones, however, pyrite grains occur along with those of other minerals. Evidently, small bits of pyrite were eroded from exposed rocks on land, carried downstream by rivers, and deposited in coastal environments, all without encountering much oxygen gas.

FIGURE 9.2. Proterozoic red beds exposed in Montana. The red color (dark here) is imparted by tiny grains of oxidized iron mixed among sand grains. Red beds are common after the Great Oxygenation Event, but not earlier.

If you want to find sandstone containing pyritic sand grains, look in deposits older than about 2.4 billion years. Other oxygen-sensitive minerals, including uraninite (UO_2) and siderite ($FeCO_3$), tell a similar tale.

Sulfur isotopes add quantitative rigor to this story. Sulfur comes in 23 isotopic varieties, all short-lived radioisotopes except for ^{32}S, ^{33}S, ^{34}S, and ^{36}S, which are stable. In chapter 8, we noted that during photosynthesis, organisms fractionate carbon isotopes, generating a geochemical signature of life that persists through time. In similar fashion, microbial sulfate reduction—anaerobic respiration by bacteria, using sulfate as an oxidant—fractionates sulfur isotopes. The degree of fractionation, recorded in the pyrite found in ancient sedimentary rocks, depends in part on the abundance of sulfate in ambient waters. And, not surprisingly, before 2.4 billion years ago, fractionation was limited, suggesting that sulfate levels in the early ocean were low. Low O_2 results in low oxidized S.

Traditionally, only ^{34}S and ^{32}S were measured by geochemists interested in the history of life and environments. Fractionation by microbes depends on isotopic mass: If we know the abundances of ^{34}S and ^{32}S in a sample, we can calculate the levels of ^{33}S and ^{36}S. That was useful, because the latter two isotopes occur in low abundances, making accurate measurement challenging. As geochemists developed techniques for measuring all four stable isotopes of sulfur, however, a surprise was in store. More than two decades ago, James Farquhar, working in the laboratory of Mark Thiemens at the University of California, San Diego, measured the abundances of all four stable isotopes of sulfur in a meteorite that originated on Mars. Surprisingly, his measurements departed from the expectation of mass dependence. Something else was going on, and Farquhar and his colleagues suggested that photochemical reactions between sulfur dioxide gas from volcanoes and high-energy solar radiation could explain the observed results—but only if the Martian atmosphere was essentially oxygen-free. Farquhar and colleagues then turned their attention to sulfur minerals deposited on the early Earth, providing a second, highly illuminating surprise. As on Mars, measured isotopic abundances in sedimentary rocks deposited before about 2.4 billion years ago displayed the isotopic pattern first observed in Martian rocks, a pattern now called mass-independent sulfur isotope fractionation. Younger samples, however, did not. Not only did this corroborate other evidence for low oxygen on the early Earth, but it also allowed early oxygen levels to be quantified: less than one-hundred-thousandth of present atmospheric O_2 levels, an extremely low number. Recently published geochemical evidence now suggests that higher (but still very low) levels of oxygen gas were sometimes present locally, if transiently, in Archean oceans. The key point, however, still stands: On the early Earth, O_2 was present only in extremely limited amounts.

All evidence, then, points toward a remarkable conclusion: 2.4–2.2 billion years ago (to include both uncertainties in timing and the likelihood that the transition was protracted), after two billion years as an essentially O_2-free planet, Earth began to accumulate oxygen gas in the atmosphere and surface ocean (figure 9.3). But why? What biological or physical changes, either alone or in combination, catapulted our planet to a new state, one with enormous evolutionary potential?

The pioneering scientists who first recognized what has come to be known as the Great Oxygenation Event, or GOE, stressed the need for the rate of oxygen production by cyanobacteria to exceed that of oxygen removal by reaction with reduced ions, gases, and minerals. On the early Earth, large amounts of reduced iron in solution emanated from seafloor hydrothermal systems, and additional reduced ions were generated by the weathering and erosion of basaltic rocks exposed on land. Also, reduced gases such as

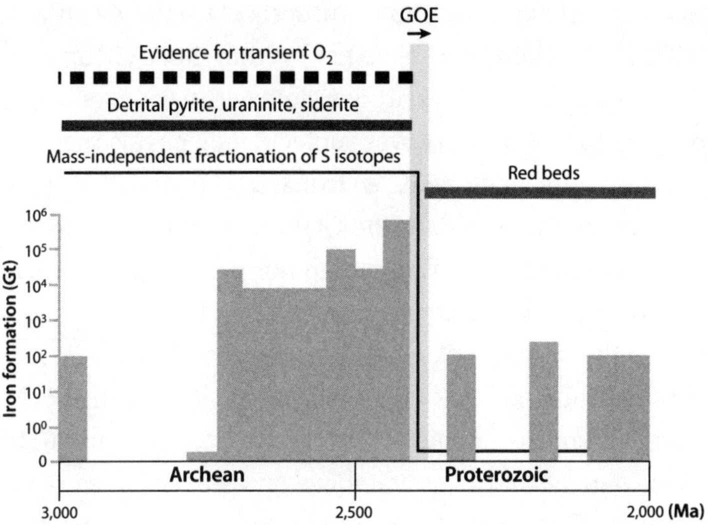

FIGURE 9.3. Summary of evidence for the Great Oxygenation Event 2.4–2.2 billion years ago. See text for discussion.

hydrogen (H_2) and hydrogen sulfide (H_2S) spewed from volcanoes. Oxygen couldn't accumulate in the atmosphere and surface ocean until the rate at which photosynthetic microbes produced O_2 exceeded the rate at which reduced materials (and microorganisms) gobbled it up. Fair enough, but what caused oxygen production to exceed oxygen consumption, and why did it occur when it did?

In many ways, the simplest hypothesis is that only about 2.4 billion years ago did bacteria evolve oxygenic photosynthesis. That would work, and it is the favored explanation of some Earth scientists. There are, however, some issues. First, there are molecular clocks—phylogenies based on molecular sequence data that are calibrated for time using fossils of known age. Most molecular-clock estimates for the origin of oxygenic photosynthesis place this event well before 2.4 billion years ago, deep within the Archean Eon. (In fairness, one or two are consistent with a younger origin of cyanobacteria.) Moreover, molecular clocks also suggest that the biomolecules needed to use O_2 in metabolism were also present long before the GOE. Evidently, oxygen gas may have been neither abundant nor persistent, but early microbes knew how to use it when present.

Turning to the rock record, Archean sedimentary successions include what are called "whiffs of oxygen," rocks whose chemistry suggests transient local accumulation of oxygen gas in the atmosphere and surface ocean. Advocates of the cyanobacteria-late hypothesis argue that such records could have formed by later percolation of oxygen-bearing fluids through older rocks—a false signal, if you will. This argument should not be dismissed lightly, but supporters of the "whiffs of oxygen" hypothesis have mounted a strong defense of the idea that these rocks document environmental conditions at their time of deposition. Moreover, multiple well-supported examples of "whiffs" are known, and if even one of them is correctly interpreted, cyanobacteria have roots well back into the Archean.

If cyanobacteria capable of generating O_2 existed long before the GOE, we must return to the argument about rates of oxygen production versus consumption. From this perspective, it isn't just the existence of Archean cyanobacteria that should concern us, but also their productivity. How much O_2 were they generating?

If cyanobacteria lived in Archean environments, their productivity could have been limited by nutrient availability, competition from other types of photosynthetic bacteria, or both. Even today, where anoxia is present in sunlit waters, anoxygenic photosynthetic bacteria—those photosynthetic bacteria that obtain their electrons from sources other than water—commonly outcompete cyanobacteria. In fact, cyanobacteria that thrive in such environments sometimes do so by shutting down part of their photosynthetic machinery, enabling them to function as anoxygenic photosynthesizers.

With this in mind, it has been argued that while the amount of photosynthesis is determined by the availability of nutrients, the type of photosynthesis, whether oxygenic or anoxygenic, reflects the availability of electron donors other than water. A number of groups have argued for low phosphorus levels in Archean oceans. This makes sense from several perspectives. First, in the water world of the Archean Eon, P supply by the weathering and erosion of rocks exposed above sea level would have been limited. At the same time, low levels of oxidants in seawater would have suppressed microbial remineralization of organic matter in the water column and sediments. (If there were little or no O_2, the concentrations of oxidants such as sulfate or nitrate ions would also have been limited—a view supported by geochemical analyses of Archean sedimentary rocks.) Consequently, microbial recycling of organic matter would have been inefficient, with nutrients in organics that fell onto the seafloor commonly staying there, rather than being freed and returned to the environment. Moreover, phosphate ions in solution adsorb onto iron oxides or react with other ions to precipitate as phosphate minerals. Thus, for reasons of both input and recycling, low P levels might well have limited total primary production in the world's oceans.

At the same time, nonwater sources of electrons for photosynthetic carbon fixation were probably relatively abundant, especially ferrous iron, as documented by banded iron formations, but also including H_2 and H_2S. As the argument goes, if the supply of these electron donors was sufficient to use up all of the P in sunlit environments, anoxygenic photosynthetic bacteria would prevail, with oxygen-producing cyanobacteria accounting for only a limited proportion of global photosynthesis. This means that cyanobacteria may well have existed long before they transformed the planet, in line with molecular clocks and "whiffs of oxygen." Only when the ratio of P of nonwater electron donors rose above a critical value would electron donors such as ferrous iron, hydrogen, and hydrogen sulfide run out before P, enabling cyanobacteria to gain global prominence. Extracting electrons from water requires a great deal of energy, but when no other electron donors are available, it is a winning strategy.

As noted earlier, toward the end of the Archean Eon, emerging continents supplied increasing levels of P to the oceans. At the same time, ferrous iron supplies from hydrothermal ridges had begun to decline as the underlying mantle cooled and, perhaps, grew less reducing. As sources of oxygen increased and sinks for O_2 decreased, the stage was set for environmental revolution. And once this transition started, positive feedbacks would have driven our planetary surface to a new steady state. Increasingly, primary production by cyanobacteria would generate a photic zone containing O_2, removing electron donors such as reduced iron and hydrogen sulfide from sunlit waters and so insuring the ecological prominence of cyanos. Pushed by tectonics, a once-alien biosphere edged closer to familiarity.

This scenario is attractive, but, as commonly happens in Earth science, it has been challenged. In a combined analysis and modeling of the mineralogy and chemistry of a banded iron formation deposited just prior to the GOE, Birger Rasmussen and his colleagues have argued that at this time P concentrations in seawater were as much as an order of magnitude higher than they are in modern seas, suggesting that P limitation was unlikely in early oceans.

The size of the earliest Proterozoic P pool has been questioned, but taking Rasmussen and colleagues' estimate at face value, how can we explain it and how does it force us to reconsider our thinking about the GOE? In the present-day oceans, P levels in shallow seawater tend to be low for the simple reason that primary producers take up this nutrient as soon as it is introduced into the photic zone. High P in earliest Proterozoic seawater could be telling us that something other than phosphorus limited primary production at that time. The obvious alternative is N—as noted earlier, even today nitrogen limits primary production over much of the oceans. That said, scientists commonly assume that on geologic timescales microbial nitrogen fixation keeps N in ready supply, highlighting the importance of P as a limiting nutrient. (There are exceptions. For many years, Paul Falkowski, one of the best biogeochemists of my generation, has argued that Earth scientists should really keep an eye on nitrogen.) Perhaps nitrogen fixation was limited by the scarcity of elements such as molybdenum, needed for the process. Or, as the generation of oxygen gas began to accelerate, perhaps microbial processes in the nitrogen cycle transformed bioavailable N compounds to N_2 gas at high rates. Or maybe the cyanobacteria-late folks were right all along.

Perhaps. But there is one more perspective that might help to explain the high P abundances measured by Rasmussen and colleagues. Today, if you want to find P concentrations an order of magnitude higher than those of surface seawater, all you have to do is dig a hole. In the porewaters of sediments just a few centimeters beneath the sediment surface, dissolved P can reach high values. Thus, the mineralogy and chemistry of the earliest Proterozoic banded iron formation studied by Rasmussen and colleagues might reflect chemical reactions that took place within sediments, rather than in overlying waters. I like this type of explanation, although if we ever converge on the true history of the GOE, it might well contain bits and pieces of several different proposals.

For now, debates about the "why" of the GOE continue, but there is no longer much argument about what happened and when. Forty percent of the way through our planet's history, a new world emerged, with extraordinary possibilities for life. In one way or another, biological processes were critical, but so were physical processes related to the emergence of large continents and mantle cooling: Earth and life in planet-changing conversation.

CHAPTER 10

The Story of O, Part 2

The Great Oxygenation Event. The GREAT Oxygenation Event. It sounds epochal, and in many ways it was. But when Earth settled into its new normal, how much O_2 actually existed in the atmosphere and oceans? Spoiler alert: not much.

In 1998, Don Canfield, then and now one of our best biogeochemists, published a thoughtful paper that set the research community on a new path regarding Earth's oxygen history. Yes, said Don, the GOE reflects an increase in O_2 production by cyanobacteria, but not much of that oxygen accumulated in air and water. More reacted with reduced minerals and gases, dramatically increasing the abundance of oxidized molecules other than oxygen gas, especially sulfate ions in seawater. In Don's hypothesis, then, the GOE resulted in a little oxygen gas and a lot more sulfate.

The discovery by Clara Blättler (now at the University of Chicago) of abundant sulfate minerals in two-billion-year-old sedimentary rocks that formed along an ancient coastline provides concrete evidence for a substantial GOE increase in seawater sulfate abundance. Score one for the Canfield hypothesis. Don's proposal, however, came with a second prediction: Although oxygen gas began to permeate Earth's atmosphere and surface oceans during the GOE, O_2

levels remained low, such that waters below the surface layer continued to be oxygen-free.

By sheer good luck, at that time that Don wrote his paper, I was beginning paleontological research on 1,700- to 1,400-million-year-old rocks from northern Australia and had accumulated an extensive collection of organic rich shales sampled from drilling projects. Don and I joined forces, with then postdoc (and now distinguished scientist) Yanan Shen doing most of the work. Shales deposited in relatively deep water showed unmistakable chemical evidence that the water immediately above the local seafloor remained oxygen-free. Consistent with Don's hypothesis, the basins we examined showed evidence of oxygen gas in surficial waters, but little or none at depth.

Buoyed by this result, my friend and colleague Malcolm Walter, Yanan, and I expanded our analysis to include a thick (about 1,000 m) succession of shales and sandstones formed 1,500–1,400 million years ago in what is now northern Australia and, once again, recovered by drilling. In this succession, features of physical geology allowed us to recognize both shallow-water, often coastal, deposits and deeper-water, offshore shales that alternated through time as sea level rose and fell. All the samples from shallow-water environments showed evidence of an oxygenated water column, whereas those from deeper settings accumulated beneath waters that were anoxic: again, oxygen in the surficial mixed layer, but little or none in the deeper ocean. What enabled us to draw these conclusions, which provided strong support for Don's hypothesis? The short answer is chemistry, and once again, it was the chemistry of iron (figure 10.1).

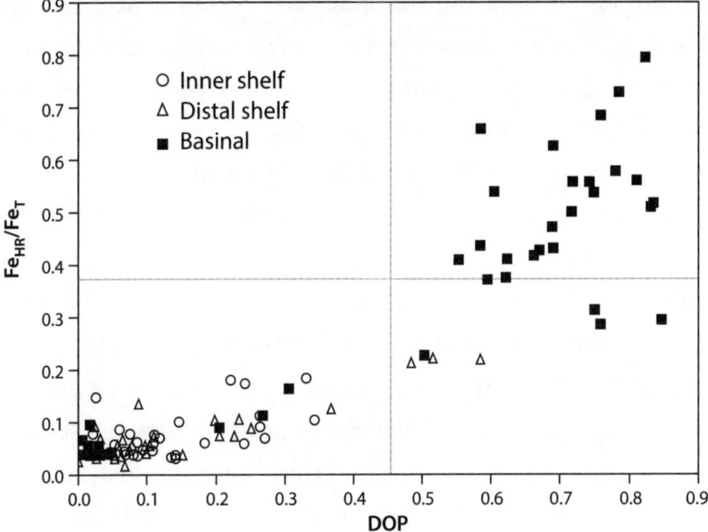

FIGURE 10.1. Early geochemical evidence that through most of the Proterozoic Eon (2,500–539 million years ago) oxygen-free deep oceans lay beneath a surface layer of moderately oxygenated water. Fe_{HR}/Fe_T stands for the ratio of highly reactive iron (largely iron in pyrite, carbonate, or oxides) to total iron (which includes iron in the minerals from which most rocks form). DOP stands for degree of pyritization, the proportion of highly reactive iron found in pyrite. In the modern Black Sea, sediments accumulating beneath shallow, oxygenated waters tend to have low Fe_{HR}/Fe_T and low DOP (lower left quadrant), just like shallow-water Proterozoic samples. In contrast, Black Sea sediments that formed in deeper environments, beneath anoxic waters, tend of have high Fe_{HR}/Fe_T and high DOP (upper right quadrant), as observed in Proterozoic samples deposited in deeper, basinal waters. See text for discussion.

If you want to examine a modern water body with O_2 at the surface but none at depth, the place to go is the Black Sea. Here, mud-rich sediments deposited beneath shallow waters differ in the composition of their iron minerals from those deposited on the deeper seafloor. The iron in mudstones from shallow environments is mostly locked within rock-forming minerals that eroded from land and accumulated on the shallow seafloor. Deep-water muds also contain iron in rock-forming minerals, but they also include iron minerals that formed within or just below the water column, especially pyrite. It turns out that iron oxides, iron carbonates, and iron sulfide minerals, collectively called "reactive iron" species by geochemists, are commonly enriched in sediments beneath anoxic waters for a reason we've discussed before. In the anoxic deep waters, reduced iron ions can be transported through the oxygen-poor waters, eventually to react with O_2, carbonate ions, or hydrogen sulfide in solution. When we analyzed the shales in our 1,500- to 1,400-million-year-old rocks, we observed the same pattern seen in the modern Black Sea—differential enrichment of highly reactive iron in sediments deposited in deeper-water settings, with a high proportion of the highly reactive iron made up of pyrite. Thus, our mid-Proterozoic rocks provided further empirical support that in the wake of the GOE, Earth remained a somewhat alien planet.

Measurements of what Earth scientists call iron speciation chemistry now number in the thousands, and they tell a consistent story. The Great Oxygenation Event may have oxygenated surface oceans, but throughout the ensuing Proterozoic Eon, subsurface oceans remained anoxic, mostly iron-rich but with local development of sulfide-rich waters. This requires that for more than 1.6 billion years after the GOE, Earth's inventory of oxygen gas remained relatively low—much lower than present-day levels.

As time went on, talented scientists like Tim Lyons and his students began to study the chemistry of other oxygen-sensitive elements in ancient sedimentary rocks. Molybdenum (Mo), for example, is highly soluble in oxic waters and can accumulate in sediments to relatively high levels; in anoxic waters, however, especially in the presence of hydrogen sulfide, it is essentially insoluble. Because of this, the amount of Mo in seawater and in sediments deposited from seawater reflects the redox properties of the waters. Its isotopic composition in sedimentary rocks also reflects oxygen availability, and both signatures corroborate the picture of moderate O_2 in surface waters but little or none at depth. Other elements, including uranium, cerium, vanadium, and thallium, also record redox conditions in ancient oceans. Even the amount of iodine incorporated into carbonate minerals provides clues to oxygen in ancient seawater. Estimates of Proterozoic oxygen derived from such data not only corroborate iron speciation data but suggest that oxygen levels were really low, perhaps as little as 1 percent or so of modern levels, or even less. Estimates of Proterozoic oxygen levels vary from one lab to the next, with some scientists pumping for higher (but still relatively low) oxygen at least at some times during the long Proterozoic Eon. While some of these varying estimates reflect differing techniques or models, most seem to be telling us that on Proterozoic Earth, levels of oxygen gas were dynamic, varying in time and space as a reflection of both local and global environmental history. Thus, while diagrams of Earth's oxygen history commonly resemble ancient ziggurats, with short times of increase separated by long intervals of invariance; there is no reason to believe that Proterozoic oxygen levels were actually constant in time or space. And we need to keep in mind the issues that confront "whiff of oxygen" studies described earlier. False signatures of oxygen increase can be imparted by oxic groundwaters long after sediment deposition.

Perhaps the best case for temporal variation in Proterozoic oxygen levels comes from rocks deposited during the GOE and in its immediate aftermath. Trace metals suggest that oxygen may have reached levels not too different from today's and that, as already noted, contemporaneous sediments include relatively abundant deposits of sulfate minerals, as predicted by Don Canfield. But such conditions didn't last. By about 1,900 million years ago, a long-lived Proterozoic state of low oxygen was in place. For all the remaining uncertainties, it seems clear that the GOE didn't result in a modern world but a prolonged intermediate state of the Earth system that gave way to our familiar oceans and atmosphere only much later.

Before considering the last great chapter in the story of O—the one that ends with a modern Earth—let's think a bit about the biological consequences of the GOE. Once the GOE established a new environmental order, it was hard to go back, and even at modest oxygen tensions, the influence on life would have been profound. As already noted, the accumulation of O_2 in shallow oceans would have removed ferrous iron, hydrogen sulfide, and other nonwater electron donors from sunlit oceans, insuring the ecological dominance of cyanobacteria. And, as O_2 began to accumulate in the atmosphere and shallow waters, microbes capable of aerobic respiration also gained ecological prominence. Aerobic respiration generates much more energy than its anaerobic counterparts, imparting a new structure to microbial communities. Moreover, as Don Canfield argued, more O_2 also meant greater abundances of other oxidants, including sulfate and nitrate ions, increasing the role of anaerobic respiration in oxygen-free environments. All in all, increasing oxidants of all types would have improved microbial reworking of organic molecules released as cells died, returning more nutrients to the water and more carbon dioxide to the atmosphere.

As oxidant concentrations increased, so did microbial chemosynthesis, the fixation of carbon dioxide into organic molecules driven by chemical reactions other than sunlight. Most chemosynthesis in modern environments takes place along the interface between oxygen-rich and oxygen-poor waters, reacting either oxygen gas or nitrate with reduced ions. The summary point is that as Proterozoic Earth history unfolded, the inventory of bacterial and archaeal metabolisms increased substantially, influencing the biological cycles of carbon, sulfur, nitrogen, and more.

For all the diversification of bacterial and archaeal metabolisms, the big biological story, halfway through our planet's history, was a new form of life: the eukaryotic cell. As noted earlier, eukaryotes don't impress with their limited metabolic repertoire; however, their shape-shifting capacity to alter cell form brought new complexity to ecosystems, as did their ability to evolve complex multicellular structures.

Comparison of extant organisms scattered across the eukaryotic Tree of Life indicates that the last common ancestor of all living eukaryotes had a genetic and cellular structure much like that of its living descendants, including a nucleus, as well as dynamic internal membrane and cytoskeletal systems, the flexible cell components that underpin the ability of eukaryotic cells to ingest other cells (facilitating the evolution of predation). Its genomic organization and regulation were also broadly like those of living eukaryotes, facilitating, in due course, the evolution of complex multicellularity. And it had a mitochondrion, the eukaryotic seat of aerobic respiration derived from a symbiotic bacterium.

While we can make some inferences about the last eukaryotic common ancestor (LECA), we know much less about FECA, the first eukaryotic common ancestor. Every year the evidence linking FECA to the recently discovered Asgard archaea increases. Asgard relatives of eukaryotes contain genes for membrane and cytoskeletal regulation that resemble those of eukaryotes, and other eukaryote-like genes are being discovered with some regularity. We still have only a limited sense of the carbon and energy metabolism of FECA, in large part because new Asgard taxa continue to be discovered, and our expanding understanding of their metabolic diversity cautions that new discoveries may well change any conclusions we might draw at present. Many biologists view the immediate Asgard sisters of eukaryotes as anaerobic, but this remains uncertain. At a minimum, the Asgard host of primordial mitochondria was probably tolerant of environmental oxygen, whether or not it used O_2 in metabolism.

Debate continues as to whether the mitochondrion was established early, within a host that was still recognizably archaean, or whether it originated later, within a host that already possessed the major features of eukaryotic cell biology. The answer isn't clear, although a recent comparison of eukaryotic genomes by Julian Vosseberg and colleagues suggests something of an intermediate solution. In their view, cytoskeletal and internal membrane complexity increased before the ancestral eukaryote lineage captured its protomitochondrial symbiont, but major gene originations and duplications that underpin modern eukaryotic cell biology postdated mitochondrial acquisition. One perspective that I've long wondered about concerns eukaryotic membranes. In bacteria and archaea, most internal membranes are stable, changing little during the cell's lifetime. In no small part, this reflects the fact that many metabolic processes are localized on membranes. With the acquisition of mitochondria and, later, chloroplasts, energy metabolism in eukaryotic cells became sequestered in the stable membranes of captured symbionts, permitting greater flexibility throughout the rest of the cell. Might this underpin the unique ability of eukaryotic cells to change shape and engulf food particles?

Mitochondria in place, LECA must have been at least a facultatively aerobic microorganism, meaning that it used O_2 in respiration when present but could probably persist anaerobically in oxygen-free environments. This, in turn, suggests that LECA postdates the GOE. Consistent with this hypothesis, Belgian paleontologist (and, I'm proud to say, Knoll Lab alumna) Emmanuelle Javaux has discovered eukaryotic fossils in about 1,750-million-year-old rocks from Australia—the oldest unambiguous fossils of eukaryotic organisms known to date. These fossils postdate the GOE by several hundred million years, suggesting that, whereas oxygen may have been necessary for eukaryotes' diversification and rise to ecological importance, it may not have been sufficient.

Various genetic and cytological innovations also constrained the timing of eukaryotic divergence. Interestingly, while early eukaryotic microfossils cannot be assigned to extant lineages with any confidence, various preserved features indicate that they possessed a flexible cytoskeletal and membrane system, enabling them to change shape during the life of a single cell, as well as genetic mechanisms to maintain cell polarity. Moreover, at least some early eukaryotes differentiated distinct cell types during their life cycle. (It is common to think that only multicellular organisms have different cell types, but many single-celled eukaryotes have distinct vegetative and reproductive cells—often cysts or spores—that alternate in time. Fossils of unambiguous cysts document this key eukaryotic feature early in eukaryotic history.) Simple multicellular eukaryotes, probably photosynthetic, indicate that multicellularity arose early in the history of the group. The key point is that one way or another, the historical union of archaea and bacteria in the wake of the GOE changed the course of biological diversity forever.

Eukaryotic fossils are relatively common in rocks younger than about 1.65 billion years, although most cannot be allied within any specific branch of the eukaryotic tree (figure 10.2). The oldest fossil eukaryotes that can be assigned to extant groups with reasonable confidence are

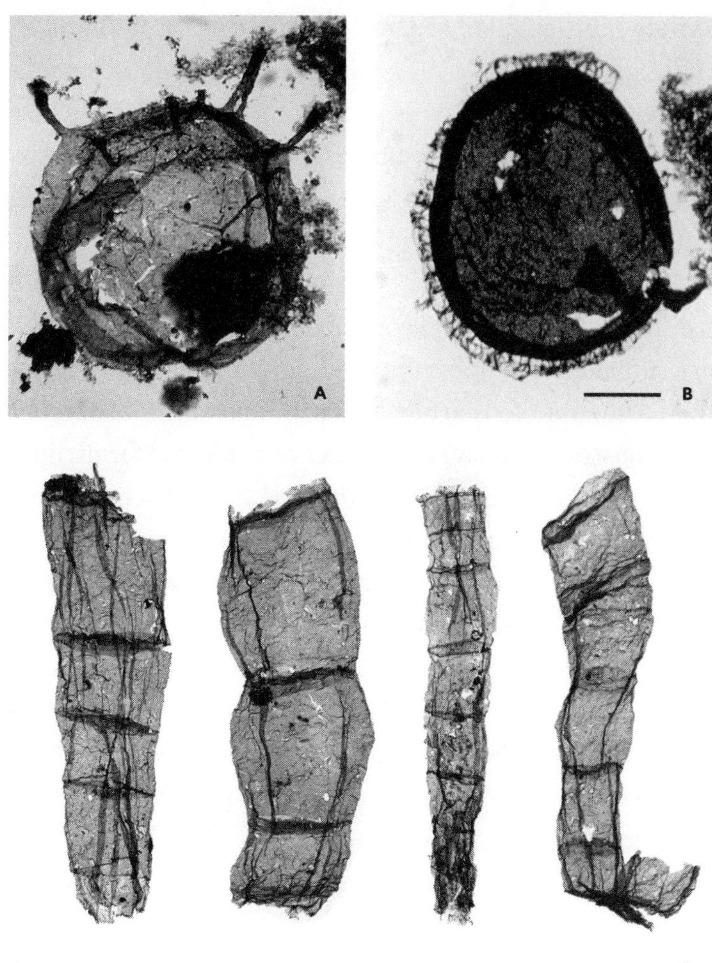

FIGURE 10.2. Fossils of early eukaryotic organisms. **(A)** *Tappania plana* (1,400–1,500 million years old). **(B)** *Shuiyousphaeridium macroreticulatum* (ca. 1,600 million years old). **(C)** *Qingshania magnifica*, a simple, probably photosynthetic, multicellular fossil (1,635 million years old). Scale bar in image B = 30 μm for A, 60 μm for B, and 50 μm for C.

simple multicellular red and green algae, both found in rocks about a billion years old. Older microfossils could belong to extant groups as well, but it is also possible that some or all of these early remains might reflect extinct forms that originated prior to the last common ancestor of all living eukaryotic organisms.

Eukaryotic microorganisms don't need much oxygen—estimated O_2 levels for Proterozoic surface environments are sufficient to support their biology—but most do need at least a bit. As noted earlier, anaerobic eukaryotes exist, but even most of these have an indirect need for O_2. Sterols, lipid molecules associated with membrane stability (cholesterol is the best-known example), require O_2 for their synthesis, and so eukaryotes living in oxygen-free environments commonly rely on their food for the sterols they need—sterols synthesized by other eukaryotes in overlying oxic habitats.

Sterols actually have a fossil record, preserving in slightly altered form in ancient rocks. Molecules derived from sterols made by green algae, red algae, fungi, and animals are present in rocks only from near the end of the Proterozoic Eon. Recently, however, a new twist on sterol history was published by Australian organic geochemist (and, again, a Knoll Lab alumnus) Jochen Brocks. Years ago, Konrad Bloch, a Nobel Prize–winning pioneer in cholesterol biochemistry, hypothesized that intermediate molecules in the biosynthetic pathway of cholesterol were once functioning structures in their own right: protosterols, if you will. In painstaking work, Jochen went looking for them in older Proterozoic rocks. And he found them in rocks as old as 1.6 billion years. Consistent with hypotheses and uncertainties attendant to morphological fossils, these protosterols could have been made by cells that predated LECA or by early offshoots of LECA and its descendants.

It appears, then, that Earth's long middle age, with some but not much oxygen in the atmosphere, was the crucible of eukaryotic evolution. Eukaryotes diversified in the oceans and in fresh water, although they remained largely unicellular or simple multicellular balls, sheets, or filaments. When did oxygen resume its climb toward the levels we breathe today, and what were its biological correlates? As we'll see in the next chapter, the third and final installment in the story of O, the rise begins about when we first see evidence of complex macroscopic organisms. How did physical and biological processes interact to shape the world we know today?

CHAPTER 11

The Story of O, Part 3

In 1959, Canadian biologist J. R. Nursall published a brief thought paper in which he ruminated over the relatively late appearance of animals in the fossil record. Recognizing even then that microbial life was far more ancient, Nursall concluded that animals—specifically, large animals likely to be preserved as fossils—could radiate only once the oceans had accumulated "a surplus of oxygen." His geological argument broadly resembled thinking about the GOE and was not complicated by actual geochemical data. Nonetheless, the idea that animal evolution reflects Earth's oxygen history has reverberated among biologists and Earth scientists since that time.

How much O_2 do animals need to function? That's not a trick question, but the answer requires a certain amount of nuance, because different animals vary markedly in their oxygen requirements. In a much-cited experiment, Dan Mills, working in the lab of Don Canfield, grew sponges in an aquarium, carefully monitoring the O_2 levels of ambient waters. Although they didn't complete a reproductive cycle, the sponges persisted at O_2 levels as low as 0.5–4 percent of the modern, leading Mills and colleagues to conclude that elevated oxygen levels were not necessary for the origin of animals. That's a reasonable conclusion, anticipated by others, including Nursall, and it helps to explain why the oldest known animal fossils are some 200 million years younger than many molecular-clock estimates for the timing of animal origins. That said, you wouldn't last long under Mills's experimental conditions, and neither would many other animal species. The statement that elevated O_2 was not necessary for animal origins is not equivalent to a statement that oxygen history is irrelevant to animal evolution.

A few years after Mills's illuminating experiment, Erik Sperling, working—fortunately for me—as a postdoctoral fellow at Harvard, gathered data from a number of ecological surveys in which biologists had reported how faunal communities change along gradients of decreasing oxygen in modern oceans. Corroborating the view that some animals can exist at low oxygen tensions, these surveys found that animals do occur at exceeding low O_2 concentrations, well below 1 percent of seawater saturation beneath the present-day atmosphere. That said, animal diversity is extremely low under these conditions and body sizes are tiny, commonly only a few hundred microns.

With increasing oxygen, both diversity and size increase, and above about 8–10 percent seawater saturation, the fauna starts looking pretty familiar. Not necessarily a happy abode for blue whales, perhaps, but sufficient for many of the invertebrate animals found on or above the seafloor. And, as oxygen availability increases, a specific functional type assumes ecological importance: carnivores. Carnivores have a relatively high oxygen demand not only because they commonly expend a great deal of energy to capture prey but also because the transient energy demand of digesting their food can be considerable. If, like many marine predators, you could stuff a prey item half your size down your gullet, the oxygen needed for digestion would be large. Carnivory is important because it provides a powerful ecological driver of animal diversification. As animals began to prey on other animals, different lineages became specialized for distinct types of locomotion and carnivory, with distinct prey preferences. Prey, in turn, became specialized in the ways that they avoid being eaten.

So the distribution of animals in the present-day ocean is consistent with the argument that increasing oxygen levels played a role in early animal diversification. And, as discussed in the preceding chapter, geological inferences about oxygen levels in the world before animals suggest that they were generally low. But when do we start to see animals in the fossil record, and what might we infer about their physiological requirements? Equally, what does the rock record suggest about the timing of renewed oxygenation, and does it resonate with the fossils?

A number of molecular-clock studies suggest that the last common ancestor of extant animals lived about 750 million years ago. Logically, that ancestor had to postdate the evolution of multicellularity in the lineage that gave rise to animals. Details of that core event are lost in the mists of history, but a combination of experiments and phylogenetic inference provides some idea of how it may have happened. In illuminating experiments, Berkeley biologist Nicole King and her colleagues have shed welcome new light on choanoflagellates, the closest protistan relatives of animals.

Choanoflagellates are generally unicellular, but King's team found that introduction of molecular signals secreted by their favored bacterial prey induced simple multicellularity in their experimental choanoflagellate population. Perhaps, the team suggested, multicellularity enhanced their ability to capture prey. Other experiments indicate that multicellularity can also enhance the ability to avoid becoming prey, as many unicellular predators can capture single-celled prey but cannot ingest cells in clusters. Thus, the early evolution of animal multicellularity might well reflect both predation and its avoidance, much as proposed for carnivorous animals and their prey as complex animals began to diversify.

Some years ago, my Brazilian colleague Dan Lahr and I proposed that the types of predator-prey relationships noted above may, themselves, have been enabled by a prior innovation, the ability of eukaryotic cells to capture and ingest other eukaryotes. The last common ancestor of all living eukaryotes was able to capture bacterial cells, and many protists feed that way today. In contrast, eukaryotes that prey on other eukaryotes—a capacity inharmoniously termed eukaryovory—are derived from bacteria-eating ancestors, and molecular-clock analysis suggested that the most prominent eukaryovores in modern oceans diverged 900–800 million years ago, not long before the dawn of animals. As inferred from the experiments noted above, we hypothesized that such predator-prey relationships among protists may have been a significant driver of multicellularity in animals and their immediate ancestors. Perhaps, perhaps not. But one way or another, function and ecology must underlie the origin of multicellularity in animals and many other eukaryotic groups.

So what do animals look like when we first see them in the fossil record? Little in your high school biology course would prepare you for the oldest fossils accepted as metazoan. Found in 576- to 539-million-year-old rocks from around the world, these creatures commonly resemble pliable air mattresses, with conjoined tubes forming sheets that spread across the seafloor or frond-like structures that waved in the currents, attached to the seafloor by stems and their holdfasts (figure 11.1). These organisms didn't have a mouth, a digestive system, gills, or appendages for locomotion. They must have obtained oxygen by diffusion from surrounding waters and fed by the action of individual cells

FIGURE 11.1. *Pteridinium simplex*, an early animal fossil preserved in late Ediacaran rocks from Namibia. Units in scale are centimeters.

on their surfaces that captured organic particles, including cells, perhaps augmented by the intake of dissolved organic molecules from the environment, much as fungi do today. At least some appear to have had muscle fibers, but their capacity to move from place to place was limited. Most metabolically active cells were in thin surficial layers; interiors of the body tubes likely contained metabolically inert fluids, a bit like the jelly in jellyfish. Given this anatomy, these early animals would have had relatively low oxygen demands.

That all sounds alien, but it isn't completely so. Placozoans, millimeter-scale animals found on the seafloor today, are little more than upper and lower cell layers that enclose a central cavity filled with fluid. They exchange gases by diffusion, feed as individual surface cells, and move about poorly, much like the earliest animal fossils, if at smaller scale. Placozoans may, thus, be telling us about the structure, function, and phylogenetic position of those fossils.

Animal fossils are relatively abundant in rocks deposited during the final 20 million years or so of the Ediacaran Period (635–539 million years ago), and indeed of the Proterozoic Eon as a whole. Well-preserved and relatively diverse fossil assemblages have been found on all continents save for Antarctica, with particularly informative deposits in Namibia, the White Sea coast of northwestern Russia, and South Australia. Most of these fossils display the same broad body plan observed for the earliest metazoans, but now some show evidence for greater complexity. *Kimberella,* for example, had a bilaterally symmetrical body, like those of most animals living today. It had a head with a mouth and a rasping organ for feeding, and it moved across the seafloor by means of a muscular foot. Trace fossils—tracks and trails preserved in sedimentary rocks—provide further evidence of complex animals capable of sustained locomotion; one type even shows evidence of leg-like appendages. And a few animals evolved the capacity to protect themselves by means of mineralized skeletons (see chapter 12), suggesting that carnivores had begun to radiate. A new world was about to begin.

That world burst into being during the ensuing Cambrian Period (539–485 million years ago). In Cambrian rocks, paleontologists find for the first time abundant animal fossils that clearly belong to still-extant groups, including arthropods, mollusks, brachiopods, and even vertebrates. Trace fossils record increasing diversity in behavior as well as morphology. Widespread mineralized skeletons document the increasing importance of both predators and prey (see chapter 12). And, for the first time, many animals moved adroitly across the seafloor or swam agilely above it. Some burrowed deep into the sediments. All in all, animals were becoming more abundant, more diverse, and more complex—and because of this, they needed more oxygen—and more food.

How did the physical world change as animals began to diversify? The chemical proxies for redox conditions noted in the previous chapter commonly show stepwise change during the Ediacaran and Cambrian periods, suggesting that the rise of animals played out in the context of increasing O_2 in the atmosphere and oceans. How much did oxygen increase, and what might explain such environmental change?

The road toward explanation is circuitous, but follow along and see where it leads. Much paleobiological attention has been lavished on the end-Proterozoic emergence of animals, but another important biological change happened on the same timescale. For most of the Proterozoic Eon, cyanobacteria were the principal primary producers in the world's oceans. Late in the eon, however, eukaryotic algae, especially planktonic green algae, rose to global prominence along the margins of continents. The evidence for this comes from fossils, but mostly fossilized molecules, rather than the morphological remains of cells. Many green algae synthesize a specific molecule called C29 stigmasterol, which preserves well in sediments in an altered but still interpretable form. As introduced in chapter 10, Jochen Brocks has meticulously studied the molecular composition of organic matter in sedimentary rocks from the mid-Proterozoic Eon into the Paleozoic. The oldest rocks known to preserve these green algal markers in abundance come from a single sample deposited near the beginning of the Ediacaran Period in Oman. By 600–550 million years ago, however, green algal biomarkers had become ubiquitous in well-preserved mudrocks around the world.

As introduced in chapter 3, in the modern oceans, water masses dominated by cyanobacteria tend to occur where nutrient availability is low. In contrast, where more nutrients are available, eukaryotic algae emerge as important primary producers. What we observe today in space may help us to understand the pattern of phytoplankton change observed in time by Brocks and his colleagues. An increase in nutrient availability would have favored the expansion of eukaryotic algae in the oceans. With more nutrients would come more photosynthesis, more food—and, yes, more oxygen.

This, of course, prompts another question. Why should nutrient levels have increased as the Ediacaran Period unfolded? Once again, tectonics enters the picture. Nearly two decades ago, the Australian geologist Ian Campbell proposed that the rise of Himalaya-scale mountains drove Ediacaran oxygen increase, facilitating, in turn, the diversification of large animals. As discussed in preceding chapters, mountains that formed during most of the Proterozoic Eon were low, limiting the rates at which their weathering and erosion could supply phosphorus to the oceans. Things changed, however, near the end of the Proterozoic Eon, as mountain building associated with formation of the supercontinent Gondwana formed high peaks not seen in more than a billion years. More recently, it has been proposed that continued cooling brought the mantle across a critical threshold beyond which higher elevations in general could be supported, marking the time from about 600 million years ago to today as different from earlier Earth history.

Tectonics, thus, may underpin Earth's tripartite history of oxygen: none at first, a bit through our planet's long middle age, and a lot in more recent times (figure 11.2). There is, however, more to the story. Ediacaran rocks preserve one more key change in environmental history. Phosphorus, generally present in the sea as phosphate ions (PO_4^{3-}), reacts with calcium and other ions to form minerals, especially a calcium phosphate mineral called apatite. These minerals can accumulate in sediments, giving rise to the commercial phosphate deposits found in fertilizers used around

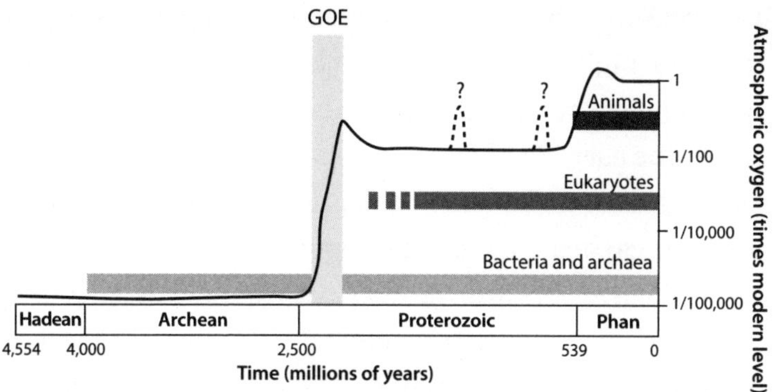

FIGURE 11.2. Broadly speaking, Earth history can be divided into three great chapters. First was a long, early interval during which O_2 accumulated only locally and transiently in the atmosphere and oceans. Following the Great Oxygenation Event, oxygen gas became a permanent component of the atmosphere and surface ocean, although most subsurface marine waters remained anoxic. Atmospheric oxygen abundances hovered around one to a few percent of present-day levels, although higher levels may have occurred transiently. Near the end of the Proterozoic Eon, O_2 began to rise again, slowly at first but achieving near-modern levels by about 400 million years ago.

the world. Relatively small amounts of phosphate minerals are found in sedimentary rocks deposited throughout the Proterozoic Eon, but in the Ediacaran Period we begin to see unprecedented amounts of sedimentary phosphate. For example, the Doushantuo Formation, in southern China, contains more phosphate ore than all older rocks combined. With the Doushantuo and several other formations marking the starting gun, large phosphatic ore deposits have continued to be deposited episodically over the past 600 million years.

In today's oceans, phosphate-rich sediments accumulate on continental shelves where upwelling—the return of deep waters to the surface—occurs prominently. There is no reason to believe that upwelling as a process began only 600 million years ago, so perhaps the nutrient load of upwelling waters increased at this time. We noted earlier that on the early Earth, when O_2 was scarce, other oxidants such as sulfate and nitrate were as well. Under these conditions, the capacity to respire organic matter that sank from the surficial waters into the deep sea was limited. In consequence, much of the P transported to the subsurface as a constituent of organic remains stayed there. My colleagues Tom Laakso and Dave Johnston and I reasoned that the rise of high mountains during the Ediacaran Period would have supplied more P to seawater, increasing primary production. Weathering and erosion of these mountains would also have oxidized pyrite in uplifted rocks, adding sulfate ions to seawater. More oxidants would have increased the efficiency with which P in organic materials transported to the deep sea would be returned to the surface, further increasing primary production.

A positive feedback would have ensued, potentiated by those great Ediacaran mountains. More P led to more primary production and, hence, more oxygen and other oxidants, resulting in more efficient remineralization of organic matter, which increased the return of P to the surface via upwelling—which led to still more primary production, oxygen, P remineralization, and so on, eventually establishing a new steady state of the Earth system. Consistent with this scenario, molecular clocks suggest that the main lineages of nitrogen-fixing cyanobacteria in the present-day oceans emerged at this time. Perhaps increasing P availability favored cyanobacteria able to supply their own nitrogen.

More primary production resulted in higher O_2 levels, but also more food, further paving the way for large animals with high metabolic demands. (Molecular biomarkers extracted from the gut of *Kimberella*, the first animal known to have had bilateral symmetry and active locomotion, show that it fed on green algae.) As animals evolved the capacity to move across and through sediments (figure 11.3), they irrigated accumulating sediments, allowing more oxygen to penetrate below the seafloor and so altering the sediment microbiota as well as the cycles of carbon and other biologically important elements.

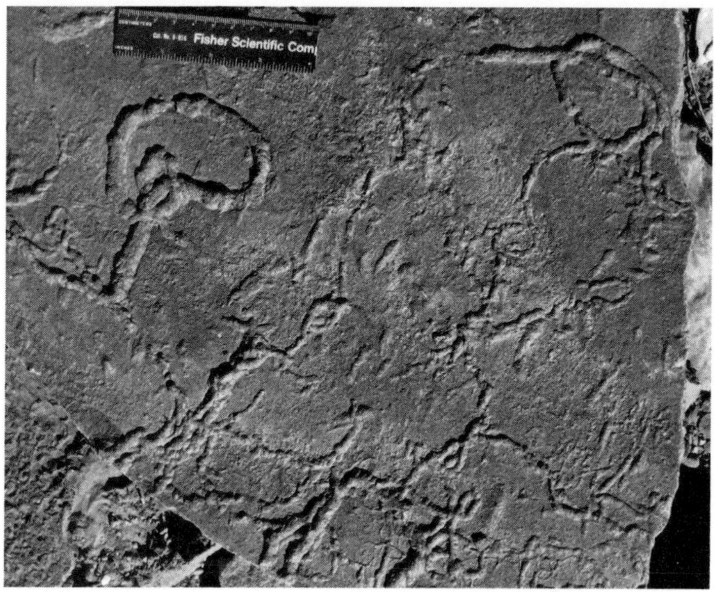

FIGURE 11.3. Trace fossils, tracks and shallow burrows made by worm-like animals at the very beginning of the Cambrian Period (ca. 539 million years ago). The image, from Namibia, shows the lower surface of a bed, so the traces are outlined in negative relief—that is, the worms moved slightly beneath the sediment surface.

A variety of chemical indicators support the hypothesis of increasing levels of oxygen gas at the time of early animal diversification, although modern levels of O_2 in the atmosphere and oceans didn't appear until later, as plants spread across the land surface (see chapter 13). Not surprisingly for a rapidly developing field, some researchers would tell this story differently, but, however simplistic, the broad picture of a planetary history with three major chapters—no, a little, and a lot of O_2—provides a sound framework for considering how the conversation between Earth and life has matured through time. Much remains to be learned, but it has become clear that we can't understand the history of life without embedding it in the story of O. And vice versa.

CHAPTER 12

The Minerals of Life

Animal, vegetable, or mineral? To Carl Linnaeus, the renowned eighteenth-century naturalist, these were the three forms of matter. For many of us, however, animal, vegetable, mineral is best known as a children's game with simple answers. A clam is animal, grass is vegetable, and a quartz crystal is mineral. Makes sense, but digging a bit deeper, how do we classify the clam's shell? Clam shells are made of the calcium carbonate minerals calcite and aragonite, set within a framework of organic molecules. These are minerals to be sure, but minerals precipitated by organisms. Scientists call them *biominerals*, at once animal and mineral.

I've long been fascinated by biominerals, not least because mineralized skeletons account for most of the animal fossil record. Many different types of organism precipitate minerals, but the formation of bones, shells, and other mineralized structures is expensive, requiring a great deal of energy that otherwise might go toward reproduction. This being the case, mineralized structures must play significant functional roles in the organisms that make them, and the benefits of investing in biominerals must outweigh the costs. Biominerals in all their diversity are fundamental to our understanding of evolution, and they play key roles in the formation of sediments on the seafloor and, so, the modern cycles of carbon, silica, and phosphorus. Not surprisingly, then, structures that are both biological and mineral participate actively in the conversation between Earth and life.

Organisms bring about mineral formation in two distinct ways: either by inducing precipitation through physiological influence on their ambient environment, or by precipitation on or within cells and tissues under direct genetic control. In sediment porewaters—H_2O in the tiny spaces between grains—metabolic processes can influence local water chemistry, either facilitating or retarding mineral precipitation. Aerobic respiration, for example, releases CO_2 into ambient waters, driving local pH down and so promoting the dissolution of calcium carbonate minerals. In contrast, in oxygen-poor porewaters, anaerobic respiration generates bicarbonate ions (HCO_3^-) that increase local pH and so facilitate calcium carbonate precipitation. Studies of ancient limestones indicate that biological induction of calcium carbonate precipitation has occurred throughout Earth's history, promoting formation of the carbonate cements that turn lime sands into limestone and microbial mats into stromatolites (figure 12.1). The main players are bacteria, especially those that respire using sulfate or oxidized iron. Bacteria that respire using sulfate also promote precipitation of another mineral: pyrite (FeS_2), formed when the sulfide ions they generate react with reduced iron present locally.

While biologically induced precipitation occurs in a wide variety of environments, most carbonates precipitated today reflect biologically controlled precipitation. That is to say, in the modern ocean most calcium carbonate sedi-

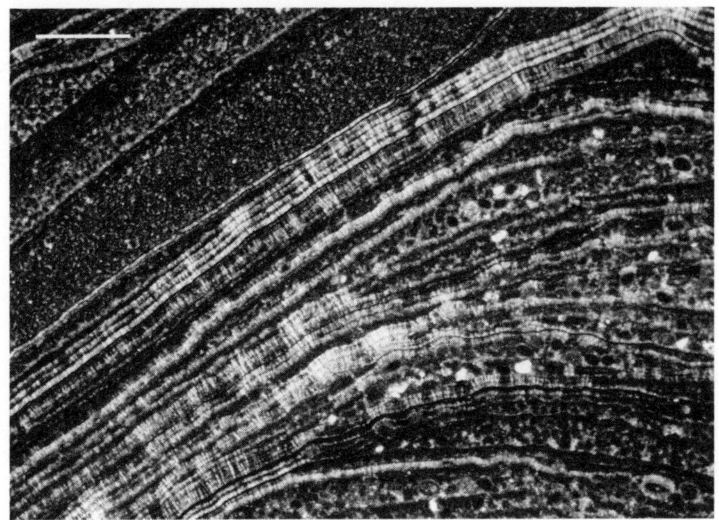

FIGURE 12.1. Biologically induced carbonate precipitation. In this 1,400- to 1,500-million-year-old carbonate rock, tiny fans of carbonate crystals precipitated on decaying microbial mats, induced by microbial processes, especially anaerobic respiration. Scale bar = 600 μm.

ments form from skeletons that accumulate on the seafloor after their makers die. Biologically induced and biologically controlled precipitation both depend on processes that increase pH and the local concentrations of calcium and carbonates ions to the point where precipitation occurs. But, while this takes place outside of participating organisms in biological induction, biologically controlled minerals precipitate in specialized spaces within cells or tissues, their chemistry modulated by physiological processes. In single cells that form carbonate tests or scales, the "privileged spaces" within which $CaCO_3$ precipitates are intracellular vesicles bounded by a lipid membrane. In animals, privileged spaces are generally fluid-filled pockets lined by cells.

Carbonate skeletons have precise and complex structures, and years of careful experimentation have shown that the mineralogy, morphology, and orientation of the crystals that make up these skeletons are determined by proteins and other organic molecules that guide biomineralization. Moreover, in many animals, the mineralized skeletons grow and change along with the bodies of their makers, requiring that skeleton precipitation be integrated within an overall program of biological development under genetic control.

If you've ever strolled along the shores of northeastern North America, you've probably seen and possibly eaten littleneck or steamer clams, piled high on plates in coastal restaurants. The clams' biomineralization is obvious in the two calcium carbonate shells that close tightly around their living tissues. As is true for many animal skeletons, the shells provide protection against predators. In these clams, however, they also serve another purpose. Littlenecks and steamers live within sediments, obtaining both food and oxygen by means of organic siphons that take in water from the overlying water column. To bury themselves in sediment, clams plunge their muscular foot into the substrate and then rock their bodies back and forth as their foot muscles contract. With this rocking motion, the shell helps the clams to saw their way downward into the sand. Many clams burrow into sediments, and experts can tell at a glance where and how different species lived on the basis of their shell morphology—even in the fossil record.

If that is why littlenecks and steamers make mineralized skeletons, how do they do it? In clams, the privileged space in which biomineralization takes place lies at the margins of a tissue called the mantle, two flaps that surround the organs for food capture and digestion, gas exchange, movement, and reproduction. Among other functions, the mantle secretes a protein covering called the periostracum, which effectively separates living tissues from the environment. The privileged space where biomineralization takes place is bounded on one side by tissues along the mantle's edge and on the other by the periostracum. Mantle cells pump both ions and enzymes into fluids within the privileged space, providing an organic framework for mineral precipitation, as well as the required calcium and carbonate ions. For years, scientists assumed that calcium carbonate precipitation in skeletons worked much like the crystal growth observed by physical chemists, one ion at a time. The truth, however, is more interesting. In many, perhaps most, organisms, $CaCO_3$ initially precipitates as tiny balls of amorphous (noncrystalline) calcium carbonate that attach to a crystal face and eventually transform into crystalline calcite or aragonite, the two mineral forms of $CaCO_3$ commonly found in skeletons. Littleneck and steamer skeletons include both an outer prismatic layer of elongated crystals oriented perpendicular to the shell surface and an inner layer called nacre (or mother-of-pearl), made up of crystalline tablets oriented parallel to the shell surface. Mediated by organic molecules, this structural organization adds greatly to the mechanical strength of the shell.

Calcium carbonate skeletons occur in many different groups of animals, including some sponges, corals, brachiopods, sea urchins and other echinoderms, mollusks, marine annelid worms, and tunicates. Carbonate biomineralization has even been observed in the organic exoskeletons of some ants. Many of these groups do not share a common ancestor that, itself, precipitated calcium carbonate, prompting the widely accepted view that carbonate biomineralization and its many skeletal products evolved independently as many as several dozen times within the animal kingdom. A few genes appear to have been co-opted repeatedly for biomineralization, especially genes involved in the cellular regulation of calcium and inorganic carbon. These genes were probably present in the last common ancestor of all animals, if not all eukaryotic organisms. Most of the genes involved in the formation of carbonate skeletons, however, are not widely shared, underscoring the independent evolutionary trajectories of different carbonate-precipitating animals.

Carbonate biomineralization is not restricted to animals. A number of red, green, and (rarely) brown seaweeds form carbonate skeletons, readily observable in coastal waters that lap onto the Bahamas and other tropical islands. Several groups of single-celled protists precipitate carbonate minerals as well, especially foraminiferans and coccolithophorids, protozoans and algae, respectively, that together account for about half of all carbonate sediments on the present-day seafloor. (The White Cliffs of Dover consist largely of coccolithophorid scales deposited 100–90 million years ago.) And, surprisingly, controlled $CaCO_3$ precipitation has even been documented in cyanobacteria. Karim Benzerara, an innovative geobiologist in France, has documented little nodules of amorphous calcium carbonate within cyanobacterial cells; their function remains unclear.

Calcium carbonate is the most widespread material found in animal skeletons, reflecting both the chemistry of seawater and the evolutionary antiquity of calcium and inorganic carbon modulation in basic cell function. It is not, however, the only skeletal mineral formed by animals. Just look in the mirror. Your bones have two major components: a calcium phosphate mineral called hydroxyapatite [$Ca_{10}(PO_4)_6(OH)_2$] and organic molecules (mostly a protein called collagen) in subequal proportions. (By volume, bones also contain about 35 percent water.) Calcium and phosphate ions come from the food you eat, and, as in carbonate skeletons, they are transported into a privileged space where they are deposited onto an organic matrix. While bones of the cranium protect your brain against injury, the internal skeleton in your body as a whole is more closely tied to locomotion, providing a framework for the body's musculature. The "choice" of calcium phosphate by humans and other vertebrate animals likely reflects their reliance on diet as a source of ions for biomineralization, as well as the relative ease of forming and reforming bones as an individual grows. Beyond vertebrates, phosphate biomineralization is not all that common among animals, occurring in the shells of some brachiopods and a few barnacles, in the mandibular teeth of many crustaceans (but only rarely in their mineralized exoskeletons, which are largely made of calcium carbonate), as amorphous phosphate granules in the stinging bristles of some marine worms, and in the skeletons of a few extinct taxa. Phosphate biomineralization is even less common in protists. Years ago, a single-celled green alga was reported to contain calcium phosphate, but the observation may actually record phosphate storage bodies (see below). Among living protists, that's about it.

The other main constituent of biomineralized skeletons is silica (SiO_2). Some sponges fashion spicules of amorphous silica, but, otherwise, silica is at best a minor feature of animal biomineralization. In protists, on the other hand, the formation of siliceous scales and tests is common—indeed, more common in terms of phylogenetic distribution than calcium carbonate minerals. Radiolaria and diatoms are particularly important in the marine silica cycle (figure 12.2). Radiolaria, among the most diverse of marine protozoans, form complex skeletons of amorphous silica, fashioning shapes that range from starbursts to micro–Eiffel Towers. These structures serve to orient cells within

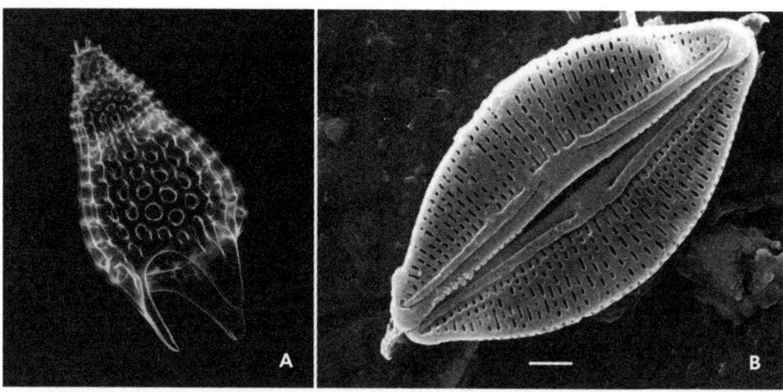

FIGURE 12.2. Silica biomineralization by radiolarians **(A)** and diatoms **(B)**. In the modern oceans, diatoms, among the world's most important primary producers, hold silica abundance within the photic zone of the oceans to exceedingly low values. Scale bar in image B = 2 µm for B and 40 µm for A.

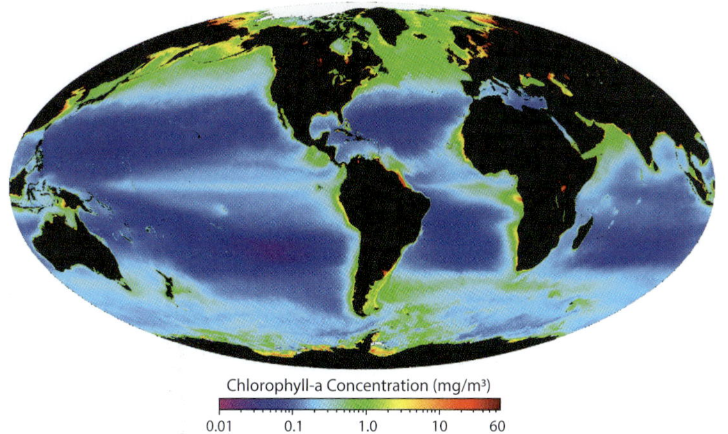

PLATE 1. The geographic distribution of primary production in the oceans, based on satellite measurements of chlorophyll content in surface waters. The bright yellow to red areas, largely along coastlines, denote places with high rates of primary production; the darkest regions, largely in the central gyres of oceans, show where primary production is low.

PLATE 2. A stromatolite, its distinctive shape reflecting the interaction of physical processes with microbial communities, from late Archean (ca. 2.7-billion-year-old) carbonate rocks in Australia.

PLATE 3. Iron formation in Dales Gorge, Australia. The 2.5-billion-year-old rocks consist of interlayered silica and iron minerals, deposited on an ancient seafloor beneath oceans far different from those at present.

PLATE 4. Proterozoic red beds exposed in Montana. The red color is imparted by tiny grains of oxidized iron mixed among sand grains. Red beds are common after the Great Oxygenation Event, but not earlier.

PLATE 5. *Pteridinium simplex*, an early animal fossil preserved in late Ediacaran rocks from Namibia. Units in scale are centimeters.

PLATE 6. Trace fossils, tracks and shallow burrows made by worm-like animals at the very beginning of the Cambrian Period (ca. 539 million years ago). The image, from Namibia, shows the lower surface of a bed, so the traces are outlined in negative relief—that is, the worms moved slightly beneath the sediment surface.

PLATE 7. Molar tooth structures in 1,400- to 1,500-million-year-old carbonate rocks from Montana. Soon after deposition, cracks developed in carbonate sediments, likely driven by gas pressure from respiring bacteria. Almost immediately, $CaCO_3$ cements filled the cracks, preserving them for all time. Molar tooth fabrics are common in Proterozoic carbonate rocks but almost unknown in younger deposits.

PLATE 8. Early evidence of biomineralized skeletons, from Ediacaran rocks. **(A)** *Cloudina,* a tubular skeleton mineralized by calcium carbonate and found widely in late Ediacaran rocks. **(B)** A population of the chalice-like organism *Namacalathus*, distributed across a bedding surface in Namibia. Scale bar in image A = 1 mm for A and 3 cm for B.

PLATE 9. Ordovician limestone found near Cincinnati, Ohio. In this and in younger limestones, skeletons are major contributors to calcium carbonate deposition.

PLATE 10. A meandering stream. Physical and biological processes combine to determine the course of rivers on the modern Earth.

PLATE 11. A late Proterozoic tillite in Namibia, its poorly sorted materials transported by ice.

PLATE 12. (A) An extensive microbial mat community in the Turks and Caicos Islands. **(B)** As shown in the mat cross section, the surface layer is tinted green by cyanobacteria, along with heterotrophic organisms that use O_2 for respiration. The purplish layer beneath it also contains photosynthetic bacteria and heterotrophic microbes, but in this case, microbes that neither produce nor utilize oxygen.

PLATE 13. Sedimentary rocks exposed in the wall of Eagle Crater, in Meridiani Planum, Mars. Note the thin layers of sandstone and the centimeter-scale concretions, colloquially named "blueberries" for their striking hue in false color images. The concretions document oxidizing groundwater that percolated through Meridiani sediments soon after they formed. Concretions can be seen here both within the sandstone and weathered out onto the Martian surface.

PLATE 14. (A) Ancient sedimentary rocks exposed in the wall of Endurance Crater, Meridiani Planum, Mars. High-angled layers in lowermost beds (arrow) reflect migrating sand dunes. The flat-lying beds above them also document sand migration across the surface. Not easily observed in this image, the uppermost beds found locally document playa lakes that formed transiently at this time. **(B)** Similar physical features can be seen in modern arid environments where sand dunes and playa lakes form, as shown here in the Namib Desert, Namibia.

PLATE 15. Europa, an astrobiologically interesting moon of Jupiter, imaged by NASA's *Galileo* spacecraft. Cracks in Europa's icy surface enable communication between the moon's surface and the ocean below.

PLATE 16. Enceladus, a moon of Saturn. **(A)** shows the cratered northern hemisphere and the "tiger stripes" that mark the southern hemisphere, from which **(B)** geyser-like fountains spew.

the water column, facilitating feeding, and they may also help to defend against protistan predators. Diatoms are among the world's major primary producers, accounting, as noted in chapter 4, for an estimated 20 percent of all photosynthesis on Earth. Found in marine environments, fresh water, and moist soils, diatoms fashion complex skeletal jewel boxes of amorphous silica that encompass the cell, providing protection, ballast in planktonic cells, and perhaps a defense against invading viruses.

Also, many plants secrete small wafers of amorphous silica called phytoliths within surface cells. It has been proposed that phytoliths constitute an energy-efficient way of strengthening plant surfaces, and the abundance of phytoliths in grasses and sedges further suggests that they provide a defense against herbivores. Consistent with this view, the fossil record shows that as grasslands spread across the Great Plains of North America 20–5 million years ago, the teeth of regional herbivores evolved ever-higher crowns, needed where phytolith-armored grasses wear away enamel.

Skeletons may be the most obvious biomineral structures, but they are by no means the only ones. Organisms are known to precipitate at least 60 different kinds of mineral, most of them not associated with skeletons. Nonskeletal minerals found within cells and tissues serve diverse functions. Take magnetite (Fe_3O_4), for example. Its hardness finds use in the radulae (tooth-like feeding structures) of mollusks called chitons, helping them to scrape food from rocky surfaces. Magnetite's magnetic properties also underpin magnetotaxis (movement oriented with respect to a magnetic field) in bacteria and navigation in a variety of protists and animals, including carrier pigeons. Magnetite has even been reported from the human brain, although its function in our own bodies remains uncertain.

Internal minerals can also help organisms to sense gravity, guiding orientation and directed growth—for example, the extension of roots downward into the soil. The minerals barite ($BaSO_4$) and celestite ($SrSO_4$) have a high density relative to that of the cells that encompass them and so occur widely in protists known to perceive gravity. Minerals can also provide biochemically accessible storage sites for nutrients. In a number of species, cells contain bodies of polyphosphate, a polymer of linked phosphate ions that stores P for later use. Also, crystals of calcium oxalate have been recognized in a diversity of algae and land plants, where they occur within intracellular vesicles and appear to function in Ca regulation, defense against predation, and, in some cases, the sequestration of toxic ions. Spinach is a champion oxalate producer.

The general picture is that organisms precipitate a wide variety of minerals. A relative handful strengthen skeletons in animals, plants, and protists, whereas a broader array of intracellular precipitates serve diverse functions in organisms distributed across the Tree of Life.

Without question, then, biomineralization plays an important role in the conversation between Earth and life. And, insofar as the physiological capacity to form minerals reflects evolution, its participation in the conversation will have varied through time. Carbonate rocks illustrate this well. As already mentioned, in the modern oceans, skeletons account for most carbonate deposition on the seafloor. This doesn't, however, mean that limestones were absent before the evolution of carbonate skeletons. For many years, I conducted field research on sedimentary rocks deposited during the Proterozoic Eon, largely before the advent of mineralized skeletons. Carbonate beds are, in fact, common in rocks of this age, many of them hundreds if not thousands of meters thick. If weathering, erosion, and hydrothermal vents continually supply calcium and carbonate ions to the oceans, the product of their concentrations will increase to the point where calcium carbonate minerals precipitate. Not surprisingly, carbonates deposited before the advent of mineralized skeletons predominantly reflect coastal environments where evaporation facilitated physical precipitation. In sedimentary successions of this age, limestones deposited in deeper-water environments commonly reflect the downslope transport of carbonate muds and sands formed initially in shallower waters.

Evidence of biologically induced precipitation is widespread in Proterozoic carbonates. For years, I used to speak of the mid–Proterozoic Eon as a papier-mâché world, as carbonate sands, muds, and microbial mats commonly were cemented into rock more or less as they formed. Carbonate crystals precipitated on and within the seafloor are thought to reflect anaerobic respiration below the surface of accumulating lime sediments (figure 12.1), as do quintessentially Proterozoic features called molar tooth structures. Molar tooth, named for its supposed resemblance to the grinding teeth of elephants, consists of cracks in limestones filled immediately after formation by carbonate cements (figure 12.3). Microbes degrading organic matter in the sediments are thought to have generated gas pressure that drove crack formation as well as increased porewater pH, thereby facilitating carbonate precipitation in the cracks. While molar tooth structures are common in Proterozoic carbonates, they essentially ceased to form by the end of the eon.

The oldest candidates for mineralized skeletons are small (10–30 μm) scales found in 800-million-year-old

FIGURE 12.3. Molar tooth structures in 1,400- to 1,500-million-year-old carbonate rocks from Montana. Soon after deposition, cracks developed in carbonate sediments, likely driven by gas pressure from respiring bacteria. Almost immediately, $CaCO_3$ cements filled the cracks, preserving them for all time. Molar tooth fabrics are common in Proterozoic carbonate rocks but almost unknown in younger deposits.

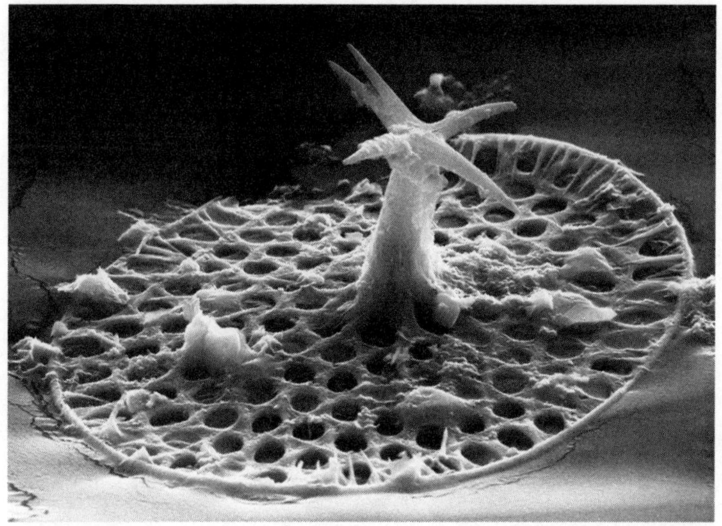

FIGURE 12.4. A phosphatic scale, part of the armor of a protistan cell. At nearly 800 million years old, these may be the earliest known example of biologically controlled biomineralization. Scale = 30 μm in maximum dimension.

rocks from northwestern Canada (figure 12.4). Like a number of living protists, the makers of these structures covered themselves in a scaly armor, presumably to provide defense against protistan predators. The scales consist of calcium phosphate. In younger rocks, especially from the Cambrian Period, there is evidence that many organic and carbonate skeletons were replicated by phosphate minerals during diagenesis (chemical and physical changes that postdate deposition but precede hardening into rock), but mineralogical arguments have been raised in support of the hypothesis that these scales were mineralized in life. If correct, this stands in marked contrast to the limited role of phosphate biomineralization in living protists.

FIGURE 12.5. Early evidence of biomineralized skeletons, from Ediacaran rocks. **(A)** *Cloudina,* a tubular skeleton mineralized by calcium carbonate and found widely in late Ediacaran rocks. **(B)** A population of the chalice-like organism *Namacalathus*, distributed across a bedding surface in Namibia. Scale bar in image A = 1 mm for A and 3 cm for B.

Animal skeletons make their debut along with fossils of soft-bodied metazoans in rocks of the late Ediacaran Period, beginning about 550 million years ago (figure 12.5). Observed diversity is low, but *Cloudina*, tubular skeletons of calcium carbonate, can be found globally in rocks of this age. The biological relationships of *Cloudina* are debated, but the fossils clearly document biologically controlled mineral precipitation by animals. Despite this evolutionary breakthrough, skeletal carbonate makes up only a minor proportion of late Ediacaran limestones.

Animals that make carbonate skeletons diversified markedly during the ensuing Cambrian Period, part and parcel of the broader Cambrian explosion of animal diversity. Nearly all animal phyla that have a fossil record first appear in Cambrian rocks, and pretty much all of the major groups known to make carbonate skeletons have done so since that time. Again, however, the contribution of skeletons to carbonate accumulation remained limited. My students and I have quantified the skeletal contribution to Cambrian carbonates in several parts of the world, with consistent results. In most environments, carbonate skeletons constitute no more than a few percent of limestones as a whole. Skeletons of extinct sponges called archaeocyathids are perhaps most common, especially in reefs, but the reefs themselves are commonly microbial.

Various explanations have been proposed to explain the advent of carbonate skeletons in Ediacaran and Cambrian oceans. Early on, many focused on changing ocean chemistry, but the observation that phosphatic and silica skeletons appeared on the same timescale strongly suggests a key

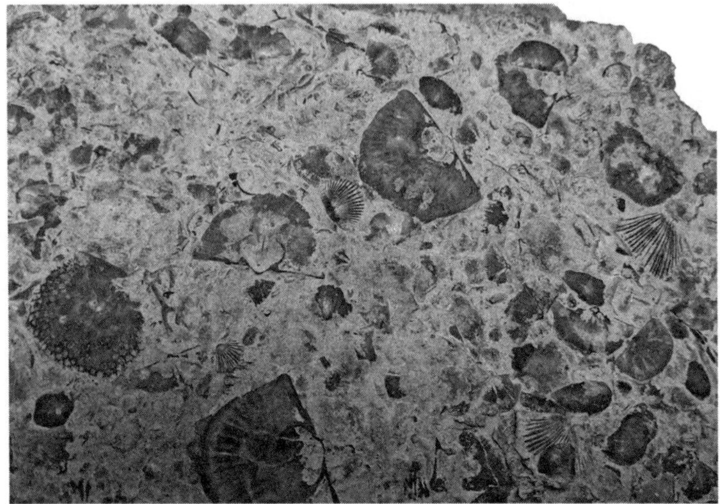

FIGURE 12.6. Ordovician limestone found near Cincinnati, Ohio. In this and in younger limestones, skeletons are major contributors to calcium carbonate deposition.

role for defense against predation. Renewed animal diversification during the Ordovician Period (485–443 million years ago) included a diversity of groups with skeletons of calcium carbonate, and since that time, limestone accumulation in the oceans has largely reflected the abundance and environmental distribution of skeletons (figure 12.6). Evolving biological participation in the silica cycle is also well documented in the rock record. Since Earth was young, silica has been supplied to lakes, rivers, and oceans by chemical weathering, as well as emissions from hydrothermal systems. Before the evolution of organisms that make silica skeletons, SiO_2 commonly left the oceans in coastal environments where evaporation drove silica saturation to levels promoting precipitation as amorphous (noncrystalline) silica, much like the case for carbonates. This silica frequently dissolved and reprecipitated locally as nodules and discontinuous beds, which in time transformed into microcrystalline quartz, commonly called chert, or flint. Black (because of organic matter inside them) chert nodules play an important role in understanding early life on Earth because they commonly preserve fossils of microbes that lived along ancient coastlines.

With the Cambrian radiation of siliceous sponges and radiolarians, silica deposition became tied to biomineralization. Cherts now formed in environments where sponges and radiolarians lived, and they no longer accumulated in and around tidal settings. Much later, as diatoms came to dominate biological silica uptake in the seas, siliceous sponges largely vacated the photic zone, retreating to the deep sea, where the dissolution of sinking radiolarian and diatom skeletons results in relatively high silica concentrations. In this regard, Manuel Maldonado, a Spanish biologist, performed an illuminating experiment. He grew siliceous sponges from Mediterranean waters in aquariums with silica levels thought to approximate those found before the radiation of diatoms. The sponges responded by forming spicule morphologies not seen in the fossil record since, you guessed it, diatoms rose to ecological prominence.

A number of scientists have tried to quantify seawater silica concentration through time. The proposed concentrations vary, but all the numbers agree that before the evolution of silica biomineralization, seawater silica levels were high; since the radiation of diatoms, silica levels in surface seawater have been exceedingly low; and in between, silica occurred at intermediate abundances. The evolution of biomineralizing organisms both mediated and responded to these long-term changes in seawater chemistry.

Why diatoms radiated only over the past 100 million years or so may reflect another global increase in nutrient availability, driven by tectonics. Some hold that silica runoff from high-standing continents helped as well. Once again, however, diatoms were not the only mineralized protists to diversify at this time. Planktonic algae called coccolithophorids, which form scales of calcium carbonate, did as well. Predation, therefore, may also have played a role. And foraminifera, heterotrophic protists that had lived on the seafloor since the Cambrian Period, diversified in the plankton at this time, spreading their calcium carbonate tests widely across the deep seafloor. Indeed, for the first time, carbonate sedimentation in the open ocean came to rival that on the shallow seafloor. Thus, in Mesozoic oceans, evolutionary innovations occurred among silica and carbonate biomineralizers, revolutionizing both plankton ecology and the geology of limestones and chert. And, once again, the drivers of this revolution included both physical and biological processes.

Animal, vegetable, or mineral? Biomineralization shows that the answers are more complicated than we (or Linnaeus) might have imagined. The voluminous formation of minerals by organisms has both influenced and been influenced by Earth's physical environment through time, providing another key theme of the conversation between Earth and life.

CHAPTER 13

Plants Enter the Conversation

Up to now, our focus has been largely on the oceans, but today most biomass, most animal diversity, and half of all primary production is found on land. What roles do the land biota play in the conversation between Earth and life? In no small part, the answers lie with one group: land plants.

The rosebush in your garden is a remarkable organism, carrying out a fundamentally aquatic metabolism surrounded by air. As introduced in chapter 5, photosynthesis takes place largely in its leaves, with carbon dioxide diffusing inward through those remarkable pores called stomata into interior cells containing chlorophyll and other parts of the photosynthetic apparatus. As noted earlier, the same pores that allow CO_2 to enter into the leaves permit water vapor to escape, potentially leading to debilitating or even fatal dehydration. To expand on the discussion in chapter 5, the rosebush combats this problem in two ways: First, the stomatal pores are lined by cells that can expand or contract based on the water status of leaf cells; at times of water stress, stomata may close, limiting water loss at the cost of decreased photosynthetic carbon fixation. Rosebushes and other plants also have a sophisticated mechanism for providing water to leaves. Roots draw water from soil and transport it upward via a vascular system composed of elongated cells that strengthen their walls with a compound called lignin. To become functional, these cells undergo programmed cell death, producing a system of porous, interconnected tubes. Evaporation from leaf surfaces literally pulls water upward through the tubes, sustaining hydration in leaves and other organs. (As noted earlier, some water is retained in cells, but, overwhelmingly, the water conducted upward by the vascular system ends up leaving the plant as water vapor.) The water-conducting cells have another important function: They provide mechanical strength for stems and roots. Proliferating conducting cells make up wood, allowing trees to grow upward toward the light while lifting leaves, fruits, and seeds out of the paths of ground-dwelling herbivores and enabling seeds to disperse more widely. Nutrients, as well as water, are taken into the plants by roots, commonly with the help of symbiotic fungi.

Thus, from the standpoint of the Earth system, we can view land plants as machines that take carbon dioxide from the atmosphere and convert it into organic molecules, mid-term reservoirs for carbon, conduits for water to move from soil to the atmosphere, and probes into the substrate in which they are anchored. What are the consequences of these processes for Earth and life?

Although the magnitude of primary production is broadly similar between land and sea, the biomass of terrestrial organisms exceeds that of their marine counterparts by two orders of magnitude. This difference in biomass is due primarily to vascular plants, which commonly live for decades or more, unlike phytoplankton, which are mostly single cells that turn over on timescales measured in days. Much of the carbon tied up in plants resides in tissues that do not photosynthesize and may not even be living (vascular tissues, as outlined above). In a mature pine tree, as much as 90 percent of its weight consists of wood and other nonliving components.

Plant cells have walls made of cellulose, and as noted above, in cells of the vascular system, the wall is reinforced by lignin, a compound not easily broken down by herbivores or detritus feeders. In consequence, organic materials derived from dead plants accumulate in many types of soil. The amount of carbon in soil organic matter is about equal to that of living biomass and atmospheric carbon dioxide combined (chapter 2). So, from the standpoint of the carbon cycle, plants remove CO_2 from air and store it as biomass and soil carbon. In swampy environments and restricted coastal waters, deposits of pure organic matter can accumulate in substantial quantities, providing the source material for coal.

A recurring theme is that the photosynthetic fixation of carbon generates oxygen, and this O_2 is kept in balance by respiring organisms that use oxygen to convert organic molecules back to CO_2. It is worth noting that photosynthetic organisms also respire, and careful observations indicate that nighttime respiration by plants can match the day's photosynthetic production, or nearly so. The plants' O_2 contribution to the surrounding environment, then, is photosynthetic oxygen production minus respiratory oxygen consumption. Similarly, a plant's net primary production is its photosynthetic carbon fixation minus its consumption by the plant itself.

When organic carbon is buried and becomes part of the sedimentary reservoir, it is no longer subject to respiration. On geologic timescales, then, it is not so much the amount of photosynthetic oxygen production that governs O_2 accumulation in the atmosphere; it is the amount of organic carbon that survives the gauntlet of respiration to become buried in sediments. It may not be surprising, then, that the final buildup of oxygen in the atmosphere and oceans occurred in and around the Devonian Period (419–359 million years ago), when plants diversified across our planetary surface. This is the time when iron chemistry first shows evidence of broadly oxygenated waters above the deep seafloor, when oxygen proxies based on cerium and molybdenum isotopes start to look modern, and when deep-sea volcanic rocks first record chemical alteration in the presence of O_2. Indeed, many Earth scientists hold that atmospheric oxygen increased to levels greater than today's during the ensuing Carboniferous Period (359–299 million years ago), aptly named for the vast quantities of coal deposited during this time. (Coal-like rocks can be found in successions as old as two billion years, but nearly all of the world's commercial coal deposits formed after the evolution of trees.) So, once again, biology plays a key role in the story of O, not necessarily because land plants photosynthesized more but because more of their tissues got buried, increasing O_2.

Of course, if the burial of plant material in sediments prevents aerobic respiration, potentially increasing the amount of oxygen in the atmosphere, it also retards the return of carbon to the atmosphere as CO_2. Because carbon dioxide is a principal greenhouse gas, increases in organic carbon burial can cool the Earth's surface. Organic carbon burial is by no means the only influence on atmospheric carbon dioxide levels, but it is probably not coincidental that the onset of Earth's greatest interval of coal accumulation coincides with a long-lived ice age, centered on the southern supercontinent of Gondwana and with a starting gun that closely follows the evolution of trees. Tectonics played a role in Carboniferous glaciation as well (see chapter 14), but most Earth scientists see a three-way linkage among plant evolution, coal formation, and climate during the late Paleozoic Era.

When I was a young scientist, two geochemical superstars sparred vociferously over questions of plants and chemical weathering. Bob Berner, an esteemed professor at Yale University, argued that the evolution of land plants increased the rate of chemical weathering across continents. Roots, he argued, increase the rate of physical weathering as they penetrate downward into soils. At the same time, a number of organic compounds released by roots increase the rate at which chemical weathering breaks down rock-forming minerals. Bacteria and fungi that respire organic compounds in the soil contribute as well. Berner supported his view by going to places like Iceland and comparing weathering on lands that did and did not have vegetation cover. Where vegetation was present, release of ions from the weathering of underlying rocks was four times greater than release from exposed rock surfaces.

John Edmond, a feisty and prodigiously talented geologist at MIT, took a different approach. Edmond spent years traveling around the world to collect water samples from the world's great rivers. He measured the amount of weathering-derived ions in his samples and found that dissolved ion load didn't correlate very well with vegetation cover. Much more strongly, it reflected the elevation of lands within watersheds that were subject to weathering. On million-year timescales, Edmond argued, rates of weathering are largely determined by what is commonly called "dynamic topography"—whether or not tectonic processes are continually raising the land surface.

Which was correct? It turns out that both scientists had made insightful observations, but with relevance on different timescales. Without question, roots penetrating into the substrate can increase both physical and chemical weathering, but if the weathered surface isn't removed by erosion, weathering becomes self-limiting. Roots can penetrate only so far; moreover, they actually stabilize their substrates, retarding rates of erosion. That is why dynamic topography is so important. Even as erosion wears down highlands, tectonic processes lift them up again, so there is always fresh rock to be weathered and eroded. In the end, the all-important processes of weathering and erosion reflect both biological and physical processes—one more example of the conversation between Earth and life.

The stabilization of soils by plants manifests itself in another, familiar way: the shapes of river channels. On the modern Earth, most river channels are straight or sinuous, their banks held in place by vegetation (figure 13.1). Sands accumulate where flow rates are low, and episodic floods

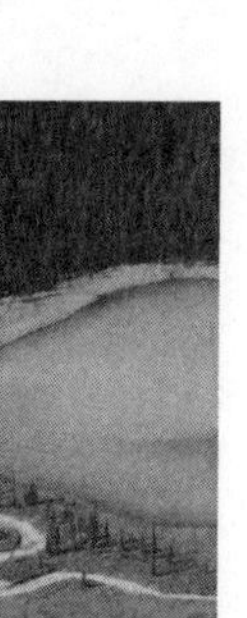

FIGURE 13.1. A meandering stream. Physical and biological processes combine to determine the course of rivers on the modern Earth.

spread mud onto adjacent floodplains. A different type of river develops, however, where sediment load is high and coarse, or where vegetation is limited. Braided rivers consist of repeatedly anastomosing channels separated by migrating bars of coarse sediment. Mud tends to get flushed downstream.

Ancient river deposits show an intriguing stratigraphic distribution. Well-defined river channels flanked by flood deposits are common from the Devonian Period onward, but uncommon in older successions. Similarly, mud is common in nonmarine sedimentary deposits formed during the Devonian Period and afterward, but uncommon in earlier successions. The difference, of course, reflects the evolution of land plants with roots capable of stabilizing channels. To be clear, the observed pattern is statistical. Braided rivers still occur today, and well-defined channel and floodplain deposits can be observed in Proterozoic strata. Sinuous river channels have even been identified on Mars, where land plants certainly played no role. In these cases, it appears that clay minerals and precipitated cements stabilized floodplain sediments. Nonetheless, the role of plants in shaping fundamental features of Earth's varied landscapes is clear.

There is another feature of the land surface that most of us encounter every day, generally ignore, and could not live without: soil. To many, soil is the product of physical and chemical weathering, fine-grained minerals that accumulate above their bedrock source. It is, however, much more than that, a premier example of physical and biological processes acting together. Soils are literally crawling with life. In addition to plant roots, there are small animals that burrow through the substrate, protists that do the same on a microscopic level, bacteria and archaea, and fungi—lots of fungi! Earthworms, nematodes, other small animals, and protists churn the soil, increasing its porosity and so aiding the percolation of groundwaters. Like roots, bacteria and fungi secrete organic compounds that bolster chemical weathering while contributing to soil's cohesion. At the same time, the myriad heterotrophic microorganisms in soils respire soil organic matter, freeing up nutrients that support continuing primary production.

Soils vary from region to region, reflecting the shape of the landscape and climate—both of which, in turn, help to determine the biota in and above the soil. The deep, dark, loamy soils in midwestern North America make possible the prodigious agriculture that feeds the continent. Similar soils in Ukraine and southwestern Russia do the same. In contrast, thin, nutrient-poor soils of arid lands support only limited plant growth. Thus, the uneven distribution of agricultural bounty reflects regional variations in this particular product of physical and biological processes.

Years ago, I had the privilege of working on the science team for NASA's Mars Exploration Rover (MER) mission. Often, as we pored over the day's new images of the Martian surface, some of my colleagues would describe fine-grained surface materials as soil. But they weren't really, at least not in the sense we understand soils on Earth. The Martian features are better classified as regolith—the products of physical and limited chemical weathering, but without the critical components that only life can provide. If you want to discover evidence of life on another planet, look for genuine soils.

The biological richness and physical diversity of terrestrial ecosystems have developed mostly over the most recent 10 percent of our planet's history. To be sure, microbial communities existed in moist areas of the land surface early in Earth history, but only as a group of green algae made their way onto land did modern terrestrial ecosystems in all their complexity begin to take shape. Those algal colonists evolved the traits that allow photosynthetic life on land one by one. First came spores with a molecular coat that retards water loss, enabling their reproductive propagules to spread across dry land. The oldest such spores are preserved in rocks from the Ordovician Period (485–444 million years ago), although a good case has been made for slightly earlier precursors. Next came the waxy coating called cuticle that prevents water loss from vegetative tissues, still in the Ordovician Period. These early land plants were simple ground-hugging organisms, at least broadly resembling modern liverworts. The ensuing Silurian Period (444–419 million years ago) witnessed the origin of erect plants with vascular tissue, the ancestors of the 400,000 or so species of vascular plants that cover the modern land surface. And then, during the Devonian Period (419–355 million years ago), plants acquired many of the features familiar to us today: large roots, leaves, wood, and seeds; many evolved more than once in different lineages. By the end of the period, Earth was a planet rich in forests, with tall canopy trees and a diversity of smaller plants in the understory.

Not surprisingly, animals invaded the land on the same timescale. Trackways show that some arthropods (the immensely diverse phylum that includes crabs, shrimp, centipedes, scorpions, spiders, and insects) ventured onto land as early as the late Cambrian or earliest Ordovician Period, about 500–480 million years ago. Whether these pioneers spent most of their time on land or ventured out of the water only transiently to feed or lay their eggs (as horseshoe crabs do today) isn't clear. By the Silurian Period, however, arthropods related to scorpions and millipedes lived beneath a canopy of air. And by the early Devonian Period, insects emerged, eventually to dominate animal diversity on Earth.

Vertebrates were relative latecomers to the party. Fossils from the second half of the Devonian Period document their acquisition through time of the morphological and physiological features needed to move, breathe, and obtain food on land. The first land vertebrates were amphibians, roaming the land surface but still bound to water for reproduction. By the Carboniferous Period, however, one lineage had evolved amniotic eggs—eggs whose shell and internal membranes defend against dehydration and predators while providing access to air and nutrients. Amniotic eggs allowed these vertebrates to reproduce on land, making them fully terrestrial. Diverse reptiles and amphibians lived among the trees, shrubs, and herbs of late Paleozoic landscapes, joined by a variety of arthropods, including insects, which began their spectacular diversification around the same time.

In fact, land animals have evolved from aquatic ancestors some two dozen times, even more if, as it seems, crabs have colonized land on multiple independent occasions. Vertebrates and several groups of arthropods clearly made the great transition, but so did mollusks (land snails), annelid worms (earthworms and leeches), and members of several lesser-known phyla (onychophorans, tardigrades, nemerteans, rotifers, gastrotrichs, and flatworms)—all derived from aquatic ancestors. Diverse protists also live in soil and on vegetation, and these similarly evolved from aquatic ancestors. Testate amoebans—amoebae that live within a flask-shaped test, or shell—provide a case in point. Their fossils are widespread in marine rocks as old as 750 million years, but today nearly all live on land. Various groups of algae, including some green algae and diatoms, also colonized the land surface, and even cyanobacteria evolved new diversity on land. Today, the most diverse branch of the cyanobacterial tree consists of morphologically complex microorganisms that live in moist soils or lake margins, a number of them in symbiotic relationships with land plants. Fossils of these diverse cyanobacteria occur in rocks containing early land plants. More generally, diverse bacteria and archaea came to thrive in terrestrial ecosystems of increasing complexity.

Fungi may be the unsung heroes of the terrestrial revolution. Soils are shot through with fungal filaments. Called mycelia, these threads weave their way through soils, seeking organic matter they can break down into simpler molecules they can absorb; fungi are particularly good at decomposing the complex molecules in plant cell walls. Mushrooms and toadstools are the most conspicuous features of these organisms, but they are simply short-lived reproductive structures; the real work goes on underground. In soils, fungal biomass dwarfs that of bacteria, archaea, and protists combined. Fungi are the principal organisms that respire organic matter and free up nutrients in soils around the world. Moreover, more than 90 percent of all vascular plant species maintain fungal symbionts in or on their roots, greatly increasing their capacity to take up water and nutrients. In the absence of this symbiosis, plants might never have gotten off (or into) the ground. Fungi also form symbioses with green algae and cyanobacteria, forming the lichens found in many landscapes. That said, fungi take as well as give; it has been estimated that pathogenic fungi that actively invade plants can decrease crop yields by some 10–23 percent. Several hundred species of fungi are even known to trap small animals in their tendril-like mycelia, qualifying them as predators. Aquatic fungi exist, but most, by far, live on land. About 150,000 species of terrestrial fungi have been described, but many biologists estimate that their true species diversity runs into the millions.

The diversity and complexity of terrestrial ecosystems has continued to grow and change since the Paleozoic Era. The main groups of extant land vertebrates—frogs, turtles, lizards, crocodiles, snakes, birds, and mammals, not to mention dinosaurs—diversified more recently, and the same is true of plants. The modern diversity of ferns and conifers evolved along with these animal groups, as did a new lineage of vascular plants: the angiosperms, or flowering plants. Today, flowering plants dominate plant diversity, supporting myriad insects, including butterflies, bees, many groups of plant-eating beetles, and more. Many insects, birds, and mammals consume angiosperm tissues, and some of them pollinate flowers and disperse fruits and seeds, ensuring the reproductive success of flowering plants.

The terrestrial biota, then, cycles water, carbon, and other elements, influencing weathering and erosion while helping to shape landscapes. Land organisms depend on climate but also help to shape it, and have much the same reciprocal relationship with soils. Indeed, on land, physical and biological processes interact in many different ways. Modern ecosystems are the legacy of the ways in which these interactions have played out over millions of years. You are part of this legacy, but it all started with plants.

CHAPTER 14

Climate and Habitability

Is Enceladus habitable? Enceladus, a small moon of Saturn, has captured the attention of planetary scientists ever since the spacecraft *Cassini* imaged active geysers that spew liquid water from beneath its icy shell. We'll dissect this question in detail in a later chapter. For now, let's ask the more general question of what habitability really means, focusing on the one planet in the universe known to be inhabited—Earth.

Earth is known to be habitable *because* it is inhabited. That much seems obvious, but it begs the question of what actually allows a planet or moon to sustain life. On our own home, liquid water is paramount, along with accessible sources of the materials found in organisms: the alphabet of elements discussed in previous chapters. In part, then, Earth's habitability reflects initial conditions, the mixture of minerals, ices, and gas that coalesced to form our planet more than four and a half billion years ago. This, in turn, reflects our position in the solar system, where most accreting materials are rocky, along with a small but vital input of water and other volatile molecules. Contrast Earth with the gas giants farther out in the solar system. It's not easy to imagine Jupiter as an abode of life.

Our solar system's distribution of rocky and icy planets reflects a temperature gradient from the hot Sun to the system's frigid outer realms. Earthlings are the lucky beneficiaries of our position in the solar system, a sweet spot where our rocky planet can maintain a reservoir of liquid water. More than one planetary scientist has christened Earth the Goldilocks planet.

From here, it starts to get interesting, especially when we note that Earth has been continuously habitable for most of its history. This merits our attention because four billion years ago, when Earth was young, the Sun's luminosity was only about 75 percent of its current value. This being the case, why wasn't our planet a frozen wasteland, rather than a cradle for life? Called the Faint Young Sun Paradox, this issue has challenged scientists for decades. Some conjecture that giant solar storms added to the radiation intercepted by our young planet, although critics counter that such events would not have provided sufficient energy to make much of a difference. Many, therefore, focus on another feature that is key to determining Earth's surface temperature: greenhouse gases. In the twenty-first century, greenhouse gases have something of a bad reputation, as carbon dioxide and other gases emitted by human activities now affect our planet in seriously concerning ways. But if Earth had no atmosphere, no greenhouse gases to warm the planet, it would still be a frozen ball locked in orbit around the Sun. On the young Earth, then, an early atmosphere rich in carbon dioxide and other greenhouse gases, including methane, water vapor, hydrogen gas, and, perhaps, nitrous oxide, played a major role in maintaining liquid water and so establishing our planet as habitable. In essence, climate made Earth habitable early in its history, and climate has sustained our planet's habitability ever since that time.

A thick atmospheric greenhouse in our planet's youth may solve the Faint Young Sun problem, but it introduces a second issue. Solar radiation has increased through time as the Sun has matured, but Earth hasn't grown too hot for life; our planet has remained in the habitable zone defined by liquid water. The long-term increase in solar luminosity, then, must be matched by a proportionate decrease in greenhouse influence. More than four decades ago, a talented trio of atmospheric modelers—Jim Walker, Paul Hayes, and Jim Kasting—proposed an ingenious solution. When we introduced the carbon cycle, we noted that weathering provides an important mechanism by which carbon dioxide is removed from the atmosphere. CO_2 reacts with water to form carbonic acid, which, in turn, reacts with rock-forming minerals. Carbon from the CO_2 becomes bicarbonate ion (HCO_3^-), which eventually reacts with calcium to precipitate as limestone. In essence, then, weathering removes carbon dioxide from the atmosphere and deposits the carbon as carbonate minerals. This process turns out to be temperature dependent. As the air warms, the reaction sequence runs faster, increasing the rate of CO_2 removal. This, in turn, works to cool the atmosphere, slowing down CO_2 removal. The feedback mechanism inherent in this relationship suggests that as increasing solar luminescence worked to warm Earth, the resulting increase in rates of CO_2 weathering counteracted this change, decreasing greenhouse capacity and so keeping Earth in its habitable state. Over the years, many scientists have added nuance to this relationship, but it still provides a broad means of understanding why Earth has remained habitable for so long.

Climate, then, is key to Earth's sustained habitability, keeping our planet in the zone where water remains liquid over much of its surface. That doesn't, however, mean that climate has been constant through time. The pas de deux between solar radiation and weathering provides first-order bumpers for climate variability on timescales of millions of years, but these are not the only influences on Earth's climate. Ocean currents transport a huge amount of heat from warm low latitudes toward the colder poles, and this physical movement of currents that carry heat poleward reflects the shapes of ocean basins, as governed by the distribution of continents. Plate tectonics, then, plays an important role in climatic variation through time. Indeed, plate tectonics plays several key roles, including the uplift of mountains, which, as we saw earlier, contributes to the rate at which CO_2 is removed by chemical weathering, as well as the storage of carbon in the mantle by subduction and the release of carbon back into the surface environment via volcanism.

Life plays a role as well, not least by removing CO_2 from the atmosphere and burying it as organic matter in sediments. Moreover, methane (CH_4), another potent greenhouse gas (see chapter 16), is produced largely by methanogenic archaea. Life also influences the albedo of the land surface. Albedo refers to the proportion of the sunlight striking a planetary (or moon) surface that is reflected back to space. Absorbed sunlight is re-emitted as heat, helping to warm the planet; reflected sunlight does not. Bare rock, deserts, and glacial ice have high albedos, meaning that they reflect a large proportion of the sunlight that strikes their surfaces. Oceans, in contrast, have low albedo, and so do forests. Thus, the rise of land plants more than 400 million years ago changed the overall albedo of our planet.

A conspicuous feature of the modern Earth is the presence of massive ice sheets around its poles. Antarctic glaciers account for the majority of our planet's fresh water, and most of the rest resides in the ice spread across Greenland. Twenty thousand years ago, glaciers were distributed more broadly, with a kilometer or more of ice covering northern North America, its southern edge running along a line connecting Long Island, central Illinois, and northern Montana. Northern Europe also lay beneath a thick blanket of ice, and sea ice around Antarctica extended much farther to the north than it does today.

So the record of glacial ice indicates that Earth's climate has changed substantially in fairly recent geologic times. Not only that, but over the past million years or so, Earth's climate has oscillated between warmer, relatively ice-poor and colder, ice-rich conditions. A century ago, a Serbian astrophysicist named Milutin Milankovitch hypothesized that this observed variation in climate reflects Earth's orbital dynamics, in particular the shape of its orbit around the sun, the angle of its axis relative to the plane of its orbit, and the direction of its axis, which wobbles through time. The discovery that climatic variations are faithfully recorded in the ice sheets of Antarctica and Greenland, and new radiometric techniques that enabled scientists to calibrate them in time, showed that, as Milankovitch had predicted, the timing of climate oscillations over that past million years reflects the periodicity of Earth's orbital variations. For this reason, we now refer to the observed climate variations as Milankovitch cycles, driven by periodic variations in how sunlight impacts the Earth, especially at high latitudes.

The earliest evidence of ice sheets in the Arctic dates back some 3.2 million years, initiating an ice age that intensified through time. Glacial ice, however, began to build on Antarctica much earlier, spreading across the polar continent during an interval of global cooling about 34 million years ago. Thus, the present ice-clad state of the planet has built over millions of years, but before the Antarctic ice sheet began to grow, there is little evidence of large-scale continental glaciation for more than 200 million years—a long-lasting, decidedly warmer Earth.

Why did Earth cool so dramatically toward the present? Most scientists point a finger at plate tectonics, which influences climate in several distinct ways. First, shifting continents can change the circulation of seawater across the planet's surface. It is no coincidence that the initial expansion of Antarctic glaciers coincides in time with the opening of the Southern Ocean. The separation of South America and Antarctica, initiating the Drake Passage, resulted in currents that swirl around the great Antarctic landmass, helping to cool the continent. Later, formation of the Isthmus of Panama three or more million years ago further influenced the poleward transport of warm, low-latitude waters, just as Arctic ice began to expand.

During the 1990s, Maureen Raymo and William Ruddiman proposed another tectonic influence on global cooling during the Cenozoic Era (66–0 million years ago), pointing to uplift of the Himalaya and Tibetan Plateau, Earth's highest landscapes. Consistent with our earlier discussion of dynamic topography, this uplift is thought to have increased rates of chemical weathering, thereby drawing down atmospheric CO_2 levels. More recently, attention has also been focused on the mountains of central New Guinea, whose rapid uplift in a tropical setting would also have contributed substantially to global cooling beginning 10–6 million years ago.

Much of the ice in eastern Antarctica melted about 16 million years ago before resuming its expansion. In fact, transient global warming is recorded in many locations around the world and attributed, at least in part, to carbon dioxide release during massive volcanism where the Columbia River flows today. Earth's climate, then, reflects the feedbacks between solar radiation and weathering, tectonic changes that shaped and reshaped ocean basins while driving the uplift of mountain chains, and biological processes that help to modulate atmospheric greenhouse gases.

Despite the geologic dictum that the present is the key to the past, a world rich in glacial ice has existed only episodically through our planet's history (figure 14.1). Two major glaciations marked the ancient continent of Gondwana during the Paleozoic Era: a long-lived ice age 350–260 million years ago, and an earlier, briefer event about 445 million years ago, with relatively warm global temperatures before, after, and in between. It is the preceding Proterozoic Eon, however,

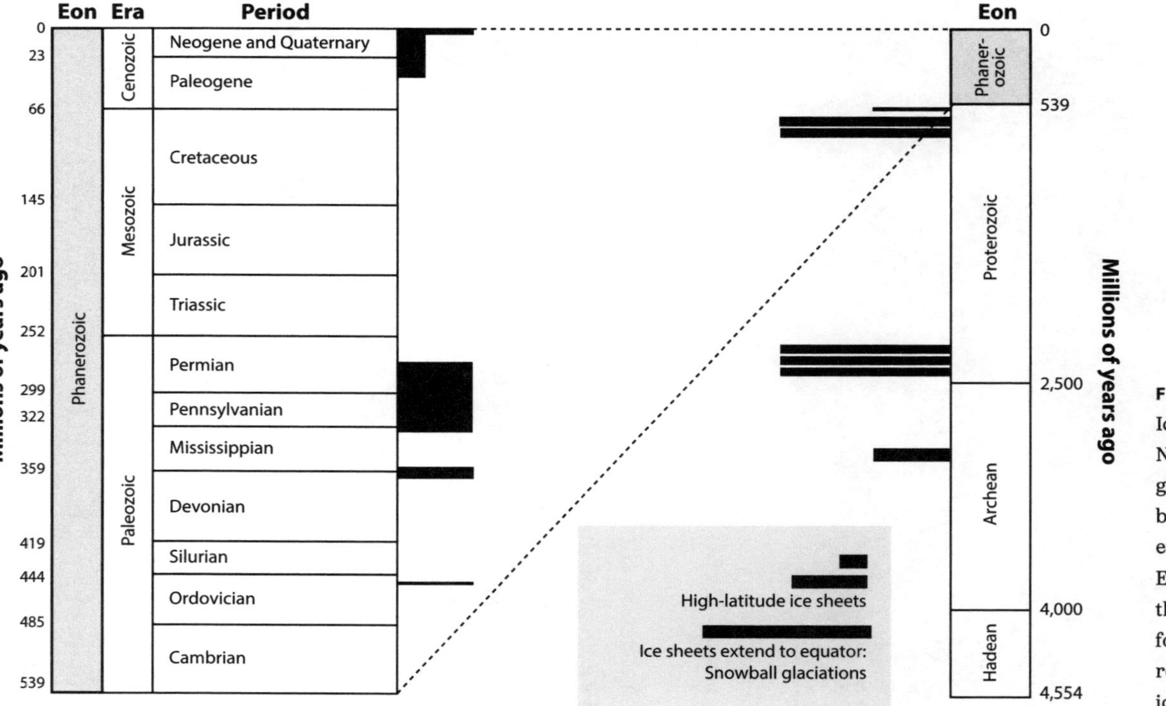

FIGURE 14.1. Ice ages through time. Note the Snowball glaciations near the beginning and the end of the Proterozoic Eon. And note as well the long intervals for which there is no record of continental ice sheets.

FIGURE 14.2. A late Proterozoic tillite in Namibia, its poorly sorted materials transported by ice.

that really shows the extremes of Earth's climatic history. For most of the eon, Earth was ice-free, a greenhouse climate lasting some 1.7 billion years. Remarkable as this may be, it is the glacial events that bookended the extended warm interval that garner most attention from Earth scientists (figure 14.2).

As early as the 1870s, a glacial origin was proposed for some Neoproterozoic sedimentary rocks from Scotland, and in the decades that followed, potentially glaciogenic rocks were described sporadically from localities on several continents. Critics doubted this interpretation, arguing that the rocks in question might reflect tectonic, rather than glacial, processes. Doubts persisted even as evidence slowly strengthened, but by the mid-twentieth century, the concept of widespread late Proterozoic glaciation began to gain traction, catalyzed by a key paper published in 1964 by W. Brian Harland, an impactful geologist at the University of Cambridge. Harland began by discussing the various lines of evidence that can identify sedimentary rocks as glaciogenic, setting out rules that skeptics were hard-pressed to refute. He then plotted the geographic distribution of Neoproterozoic glacial deposits and used this to argue that the Earth had endured a global ice age shortly before the age of animals began. Harland's hypothesis was controversial at first, and the emerging paradigm of plate tectonics, now seen as key to understanding these rocks, was used to provide an alternative explanation. Perhaps continents had sequentially moved across high latitudes over millions of years, suggesting that this time interval was notable more for its restless lands than for extreme cold. Moreover, climate models suggested that should Earth ever descend into global glaciation, it couldn't get back out again.

As more data accumulated, including paleomagnetic evidence for low-latitude ice, the hypothesis that Earth had experienced a global ice age began to gain widespread acceptance. (Magnetic minerals such as magnetite orient along Earth's magnetic field as they grow; this orientation can be preserved in the rock record, providing a means of determining the latitude at which ancient magnetite-bearing rocks formed.) Importantly, Joe Kirschvink, a creative Earth scientist at Caltech, figured out how Earth could escape from pole to equator refrigeration. Such conditions, he argued, were self-limiting. Processes that remove CO_2 from the atmosphere, mainly photosynthesis and chemical weathering, would have been severely curtailed by global ice cover, but processes that add CO_2 to the atmosphere—volcanism, in particular—continued apace. In consequence, CO_2's atmospheric greenhouse effect increased until it reached a critical level, triggering deglaciation. Paul Hoffman, a formidable field geologist, further advanced the argument by understanding that the scenario hypothesized by Kirschvink could account for unusual carbonate rocks and other features of beds immediately above glaciogenic rocks.

Indeed, glaciers did reach sea level at the equator, and climate models suggest that the oceans must have been largely or entirely veneered by sea ice—a Snowball Earth, in evocatively plain language. Moreover, increasing radiometric constraints show that it happened twice during the Neoproterozoic Era, first between 717 and 659 million years ago and again from about 639 to 635 million years ago. And Snowball conditions also preceded the long Proterozoic greenhouse, with a series of glacial advances that began 2.45 billion years ago.

The earlier Proterozoic ice age coincides with the Great Oxygenation Event, when oxygen gas first permeated the atmosphere and surface oceans, and a popular view holds that accumulating O_2 scrubbed methane and other reduced greenhouse gases from the atmosphere, taking their greenhouse warming along with it. The late Proterozoic ice ages, in turn, are sometimes ascribed to a major eruption of basaltic volcanic rocks, which weathered rapidly in low latitudes, removing carbon dioxide from the atmosphere.

The causes and consequences of Proterozoic ice ages continue to inspire debate and its scientific cousin, research. One question I find particularly interesting is why Proterozoic glaciers extended to the equator, but younger ice ages (including one or more short-lived regional events near the very end of the Proterozoic Eon) did not. Despite continuing uncertainties, however, the main point is clear: The climatic variation experienced by Earth over the last few million years does not begin to approach the extremes of heat and cold experienced over the long course of geologic history.

The preceding discussion may suggest that life has been a passive beneficiary of the physical processes that keep Earth habitable. Certainly, physical processes exert a strong influence on climate through time, but life also has a voice in the conversation. Some biological influences may seem obvious in retrospect. For example, to the extent that the Great Oxygenation Event spurred Snowball glaciation as the Proterozoic Eon began, we must consider cyanobacteria, source of the O_2 that transformed our planet's surface. Also, in an earlier chapter, we discussed how land plants, with their production of wood and other hard-to-degrade tissues, increased rates of organic carbon burial beginning in the Devonian Period, helping to usher in the late Paleozoic ice age.

Plants have other influences as well. As noted earlier, forested lands have lower albedo than bare rock, helping to warm the planetary surface. This suggests that the evolution of land plants may provide at least a partial answer to the question of why Snowball glaciations are limited to the Proterozoic Eon, before plants originated. Plants clearly reflect the global distribution of climate zones, but, again, the conversation is not completely one-sided. For example, the luxurious vegetation of tropical rain forests reflects persistently high rainfall, but there is reason to believe that rainfall patterns over the Amazon, the Congo, and Southeast Asia owe much to the flowering plants that dominate these ecosystems. Because of their anatomy, flowering plants transport water upward from the soil at high rates, much higher than those of conifers and other plants. Most of this water is returned to the atmosphere as water vapor, providing a steady supply of water for regional rainfall. Models suggest that in the absence of flowering plants, tropical climates would be hotter, drier, and more seasonal, a conclusion supported by the geological record.

Animals have important roles to play, too. As noted earlier, animals that burrow into marine sediments aerate their surroundings, increasing the rates at which carbon and nutrients are returned to the atmosphere and overlying waters. And similar things happen on land. Until recently, I had never heard the term *biopedturbation*, but Lucy Wilson, a biology graduate student, introduced it to me. Biopedturbation is the terrestrial analog of the bioturbation observed on the seafloor. A number of animal species bore or burrow into soils, facilitating the remineralization of soil organic matter, freeing carbon and nutrients much as occurs within the seafloor. Indeed, it has been estimated that the nutrients released by earthworm biopedturbation in fields can increase crop yields by as much as 25 percent. Other estimates are lower, but there is consensus that biopedturbating animals play a significant role in regulating the environment and feeding the human population.

Perhaps the most remarkable aspect of Earth's habitability is that it has lasted so long—four billion years and counting. In combination, sunlight, tectonics, ocean circulation, and organisms have populated a complex system of feedbacks, facilitations, and other interactions that have enabled life to persist through most of our planet's history. Clearly, since life became established, Earth has never veered out of the habitable zone, and there is no reason to believe that it will do so in the foreseeable future. That is not to say, however, that throughout its history, Earth has been a peaceable kingdom. In the next chapter, we'll consider a number of times in the past when perturbations in the Earth system brought life uncomfortably close to its limits.

CHAPTER 15

Close Calls?

Recently, a group of scientists argued that about 250 million years from now, continuing increases in solar luminosity and the assembly of our currently dispersed continents into a single supercontinent centered on low latitudes will result in terrestrial environments too hot for mammals, exterminating a diverse and ecologically important group of animals that includes—today, at least—humans. Other scientists have pushed back a bit on this idea, not so much denying its plausibility but, rather, claiming that mammalian endgame lies in a more distant future. I'll admit that I seldom worry about events that won't happen for many millions of years, but the argument highlights an important point: While Earth has been a biological planet for most of its long history, life has been challenged repeatedly by both long-term and transient events. Over Earth history, how close has life come to the brink? And will we ever ride over its edge?

For all we know, life could have emerged and disappeared multiple times on the young Earth, with nascent organisms unable to expand beyond local, highly vulnerable populations in volatile environments. Nonetheless, there is good reason to accept that all organisms alive today are descended from a single ancestral population that lived some four billion years ago, with countless generations not only persisting but diversifying through time. How did the Earth's surface remain in the habitable zone all this time, not too hot or too cold, too dry, or too stressed by inadequate nutrients or toxins? Parts of the answer can be found in previous chapters. Putting them together helps us to appreciate our planet as a complex and remarkable whole.

In chapter 14, we discussed how greenhouse gases have kept Earth in the habitable zone since its youth, even though solar luminosity has increased by about a third over the past four billion years. The feedback between weathering rate, which removes the prominent greenhouse gas carbon dioxide from the atmosphere, and temperature goes a long way toward explaining the guardrails that keep Earth habitable. There are, however, other players in this game. As discussed in previous chapters, a variety of physical and biological processes work together to maintain a clement planet while cycling the nutrients that sustain primary production and, hence, all life. Phosphorus, nitrogen, and other elements in buried organic matter and minerals accumulate in sediments, to be restored to the biosphere by organisms, uplift, and erosion. In the absence of plate tectonics, maintenance of Earth as a habitable planet might have been difficult, at least for a biota capable of leaving a large footprint on our home.

We don't know precisely when plate tectonics originated, but it persists because Earth's interior remains sufficiently hot to drive convection of our planet's mantle. Earth's hot interior also accounts for the molten iron in its core, which generates the magnetic shield that protects our planet from harmful radiation and the solar wind, a constant stream of particles emanating from the Sun and capable of stripping away the atmosphere through time. The question of just how much of Earth's habitability—at least for microbes—depends on the magnetic shield is contentious, but, to the extent that the shield protects us from solar wind's erosive effects on the atmosphere, we benefit greatly. It is well established that heat slowly but inexorably escapes from Earth's interior. Thus, our planet's internal temperature is cooling, and has been for billions of years. At some point, cooling temperatures could shut down the plate tectonic engine, imperiling life, but that won't happen for a long time to come.

As far as we know, Earth is unique within our solar system in the presence of plate tectonics, and so our Goldilocks status extends beyond our distance from the Sun to include our planet's size and composition, which underpin Earth's internal heat engine and, so, tectonic history. As much as the position of Earth's orbit around our long-lived Sun, it is this continuing action of physical planetary processes that underpins life's persistence through time. As I've argued many times to my students and colleagues, it may be that what makes Earth unusual is not so much that life began here—that may happen widely in the universe—but that it has persisted so long.

At least at a few times during the three billion years or more when life was mainly microbial, biological diversity may have been placed in jeopardy. It isn't easy to drive bacteria to extinction, given their immense numbers, wide distribution, and genetic versatility. Nor, given their simplicity, is it easy to identify extinct groups from bacterial microfossils. That said, comparative molecular biology offers some clues to extinct microbes. Iron metabolisms illustrate the approach and the insights it allows. As discussed in previous chapters, oceans on the early Earth were both anoxic and iron-rich. Metabolisms such as iron-based photosynthesis and iron respiration are thought to have been widespread in those seas, playing major roles in the early carbon cycle. However, when MIT scientists Erik Tamre and Greg Fournier plotted the distributions of these metabolic pathways on the Tree of Life and then estimated the times of origin for taxa capable of iron metabolism, they found that these groups originated relatively recently, hundreds of millions of years ago to be sure, but nowhere near the multibillion-year dates that would account for Archean metabolisms.

To a much greater extent than eukaryotes, bacteria and archaea obtain and utilize genes from microbes to which they are only distantly related, a process introduced earlier as horizontal gene transfer. Based on the distribution of iron metabolisms on the Tree of Life, Erik and Greg demonstrated that the genetic apparatuses needed for these capabilities have been transferred laterally from one bacterial lineage to another repeatedly through time. Their conclusions: Iron metabolic pathways originated early in Earth history and through time have spread to newer branches via horizontal gene transfer. That is, the genetic pathways for iron metabolism are ancient, but the modern bacteria that possess them are not. As far as we can tell, the iron metabolizers in Archean oceans are extinct.

Perhaps the first challenge to Earth's biota that might be inferred from the geological record was the key event that made much of life as we know it possible: the Great Oxygenation Event (GOE). As discussed in chapter 9, 2.4–2.2 billion years ago, oxygen began its permanent accumulation in the atmosphere and surface oceans. This jump-started the radiation of organisms that gain energy via aerobic respiration—organisms like us. But the oxygenation of Earth's surface ocean probably spelled doom for at least some anaerobic microbes that had long inhabited surficial environments. The late, great Lynn Margulis called the GOE "an oxygen holocaust," a poor choice of words, perhaps, but a clear expression of the view that anoxygenic photosynthetic bacteria were imperiled as oxygen gas scrubbed alternative electron donors from sunlit waters. At the same time, the infusion of O_2 strongly favored heterotrophs capable of aerobic respiration, also challenging anaerobes in surface waters. All true, and although it is impossible to quantify, the GOE may well have doomed many microbial populations, including some of the early iron-based photosynthetic bacteria mentioned in the previous section. That said, as observed in chapter 10, while the surface ocean was infused with at least a bit of O_2, most of Earth's oceans remained oxygen-free. And even today, anoxic environments are relatively widespread, including sunlit but oxygen-free environments where anoxygenic photosynthetic bacteria continue to thrive. A cross section cut through modern microbial mats on the seafloor commonly reveals a purple layer beneath the blue-green cyanobacteria at the surface. The purple hue comes from anoxygenic photosynthetic bacteria that live a few millimeters below the surface, where light penetrates, but O_2 does not (figure 15.1). The purple layer and points below it also harbor anaerobic heterotrophic microbes. That is, for microbial communities that covered sunlit seafloors for billions of years, the advent of environmental O_2 enabled cyanobacteria and aerobic heterotrophs to spread widely, but microbes with oxygen-free metabolisms persisted just beneath those surface communities. This suggests that the major consequences of the GOE for Earth's early, anaerobic microbiota may have been a change in environmental distribution, rather than wholesale extinction.

But other challenges were afoot. As noted in the last chapter, the GOE coincides with a series of globe-swaddling glaciations in which ice sheets extended to the equator and much if not all of the oceans were covered in sea ice, a Snowball Earth. Like the GOE itself, a climate so extreme would have impacted Earth's biota. Snowball climatic conditions occurred both at the beginning and near the end of the Proterozoic Eon, and as the younger occurrences are far better known, let's focus on them first and then extrapolate backward.

Many have asked how life would have been affected by Snowball glaciation. Perhaps a better question is how anything could survive such an extreme climate. Today, many different kinds of organism live in small, water-filled holes on sea-ice surfaces. Paul Hoffman, among others, has argued that such features, called cryoconites, would have sustained life through the glacial ages. Hot springs, found today in settings as various as Yellowstone Park and Iceland,

FIGURE 15.1. (A) An extensive microbial mat community in the Turks and Caicos Islands. **(B)** In the mat cross section, the surface layer is tinted green by cyanobacteria, along with heterotrophic organisms that use O_2 for respiration. The purplish layer beneath it also contains photosynthetic bacteria and heterotrophic microbes, but in this case, microbes that neither produce nor utilize oxygen.

provide other possibilities, as do lakes in local ice-free environments and, perhaps, locally ice-free spots in the oceans. We don't know in detail how life survived, but it did. Both fossils and molecular-clock estimates indicate that many still-extant groups of bacteria, archaea, and eukaryotes originated prior to the glacial events, so refuges of some type—probably multiple types—must have been available.

There is no way to know how many bacterial and archaeal populations perished during Snowball times; all we really know is that the major microbial metabolisms persisted, and so the principal biological components of Earth's carbon, sulfur, nitrogen, and other biogeochemical cycles continued to function. We're in a bit better shape for eukaryotes, as at least some of them left interpretable fossils. Fossils show that red and green algae originated well before the onset of Snowball glaciations, as did testate amoebas and other kinds of protozoans. Indeed, molecular clocks suggest that most major groups of eukaryotic organisms have preglacial histories, including animals, which could have originated as much as 750 million years ago. Of course, this doesn't mean that no extinctions occurred. Most of the protistan microfossils found in preglacial rocks do not occur in beds that overlie Snowball records. It may be that Snowball glaciations extirpated many eukaryotic species but few major groups. (A significant caveat here is that groups that disappeared entirely would not show up in phylogenies based on comparative biology, and if such groups left no interpretable fossil record, they would simply go unrecorded in geologic history.)

Glacial events near the beginning of the Proterozoic Eon are increasingly interpreted in terms similar to those inferred for their late Proterozoic counterparts. To the best of our knowledge, eukaryotes did not yet exist, and the Archean record of bacterial evolution is limited. Very likely, individual populations perished while major metabolic groups persisted, as inferred for the end of the eon. In any event, neither early nor late Proterozoic Snowballs ended life on Earth, although they may have provided some of our closest calls.

Biological catastrophes during the past 500 million years are better documented, in no small part because animals have left a rich fossil record (figure 15.2). Nearly all animal species that have ever existed are extinct, so extinction per se is common; indeed, it is the expected fate of species. Perhaps less expected is the observation that at several moments over the past 500 million years, large numbers of species disappeared at the same time. Called mass extinctions, these events underscore both the vulnerabilities and persistence of animals and other forms of life through time. Based on an extraordinary compilation of marine animal diversity through time, pioneered by the late Jack Sepkoski, we know that at five moments in the age of animals, mass extinctions changed the course of animal evolution. Interestingly, the "Big Five" do not share a common cause. The best-known extinction, 66 million years ago, when dinosaurs perished, occurred in concert with the impact of a large meteorite. A larger extinction event, some 252 million years ago, relates to massive volcanism, as does a

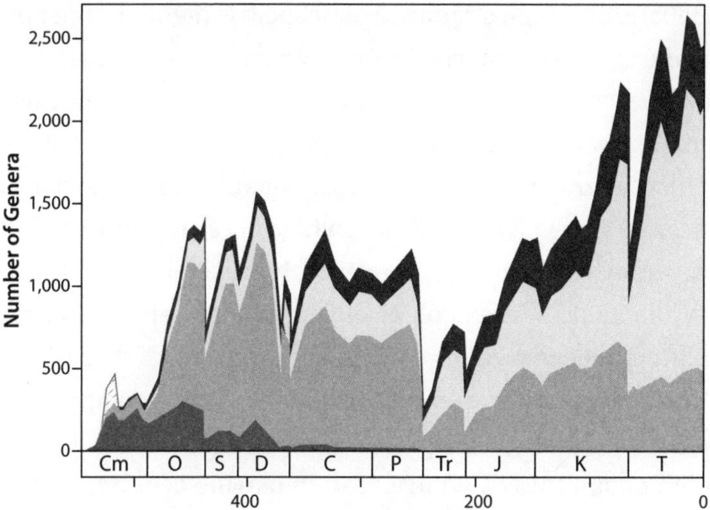

FIGURE 15.2. The diversity of marine animal genera recorded as fossils through the Phanerozoic Eon. Five moments in time record sharp drops in genus richness. Mass extinctions occurred at the end of the Ordovician Period, the later Devonian Period, the end-Permian Period, the end-Triassic Period, and the end-Cretaceous Period.

third extinction about 200 million years ago. A fourth mass extinction took place about 445 million years ago, coincident with an ice age that interrupted an interval of high sea level, eliminating extensive shallow-water environments populated by marine animals. Recent data suggest that this diversity decline began before the ice age, possibly driven by massive volcanism. The last of the Big Five is a bit different in that diversity decline is distributed through tens of millions of years, culminating about 360 million years ago. Curiously, this protracted drop in marine diversity is associated with declining speciation rates as much as concentrated extinctions. Causation remains a topic of debate, although a number of proposals highlight the episodic expansion of oxygen-free waters in the oceans.

The consequences of these events were profound; the extinction 252 million years ago is estimated to have eliminated as much as 90 percent of animal species in the oceans. And yet, the biota as a whole persisted, and animal diversity recovered. It recovered, however, in a new form. While many of the ecologically dominant animals in pre-extinction ecosystems disappeared, survivors diversified into now permissive ecosystems where limited competition facilitated biological innovation. Brachiopod- and trilobite-rich communities, widespread in marine ecosystems for more than 200 million years, collapsed, but in time, surviving mollusks, cnidarians, and arthropods generated the bivalve, gastropod, coral, and decapod biota we see today. (Decapods are an order of marine arthropods that includes lobsters, crabs, shrimp, and more.) Similarly, meteorite impact 66 million years ago may have wiped out the dinosaurs, but within a hundred thousand years or so, mammals, which had long lived in the shadow of the thunder lizards, had begun their radiation to produce the diversity found in modern terrestrial ecosystems.

The bottom line? Past mass extinctions demonstrate that complex organisms are vulnerable to perturbation originating both within the Earth system and raining down on us from space, but through it all, life has remained at some distance from total demise. In fact, the extinction of dominant species has repeatedly opened up new evolutionary possibilities for survivors.

The preceding discussion has focused resolutely on the past, when catastrophes altered but did not terminate the course of evolution. Throughout recorded history, however, most people have anticipated *future* apocalypse, imminent and ordained by God. Secular agents of planetary collapse entered the conversation only in the nineteenth century, starting, perhaps surprisingly, with George Gordon, Lord Byron, whose aptly named poem "Darkness" envisioned a world destroyed by a faltering sun, immense volcanoes, and the like. This may resonate with geologic history (unknown at the time), but as related by Dorian Lynskey in his entertaining and illuminating *Everything Must Go: The Stories We Tell About the End of the World*, Byron was inspired by a contemporary event: the year without a summer. In 1815, the devastating eruption of Mount Tambora, an Indonesian volcano, spewed huge amounts of ash and sulfate aerosols into the atmosphere, causing a global drop in temperature. In consequence, the following year saw widespread crop failure and famine in the Northern Hemisphere. From this point onward, it became apparent that the end of the world might be inherent in the world itself.

More recently, in the wake of the devastation at Hiroshima and Nagasaki, nuclear holocaust has emerged as a favored cause of apocalypse, for humans if not for all living things. Hyperintelligent machines, plague, climate catastrophe, and zombies also have their adherents, if nothing else underscoring that writers, filmmakers, and many of the rest of us are drawn to cataclysms.

In plain-spoken New England verse, Robert Frost once considered whether the world will end in fire or ice. The geologic record indicates that, over Earth's long history, both have repeatedly devastated organisms on land and in the sea, strong evidence for the view that catastrophe is sewn into the fabric of our planet and the solar system around us. Nonetheless, life has persisted, and so, if we are fortunate enough to avoid nuclear catastrophe, Earth may remain a biological planet for a long time to come. Plant and animal species will come and go, buffeted on rare occasions by large meteorites or giant volcanoes, and both groups will face their reckoning in the distant future, as introduced at the beginning of this chapter. That said, estimates of life's final curtain vary quite a bit. One widely discussed proposal holds that as solar luminosity continues to increase and atmospheric CO_2 correspondingly decreases, carbon dioxide levels will fall to levels below those that support photosynthesis, terminating life a billion or so years from now. Such models come with large uncertainties, but it does seem likely that by 7.5 billion years in the future, when the Sun inevitably expands to become a red giant that consumes Earth, our planet will long since have returned to the lifeless state of its origin.

CHAPTER 16

The Conversation in the Twenty-First Century

When I first began to draft this chapter, in the spring of 2024, I opened with the news of the day: 2023 had just been declared the warmest year ever recorded, perhaps the warmest experienced by humans in some 125,000 years. With unwitting prescience, however, I followed that statement with "at least until 2024." By the following spring, as I worked to finalize this volume, 2024 had eclipsed the previous year as the warmest on record. And January 2025 established still another record, at least for that month. By year's end, 2025 may not emerge as the new record holder, but it will certainly rank among the 10 warmest years since people began to record temperatures.

The years 2023 and 2024 didn't just inch beyond previous records; they shattered them. And it isn't simply average temperatures we need to consider. With increasing warmth comes an increase in extreme weather, dangerously so in many parts of the world. In 2023 and 2024, major wildfires scorched all continents. North Americans will particularly remember dry forests that burned across Canada, spreading smoke across much of the continent. (If we needed a reminder, tinder-dry woodlands in northern Canada ignited again in 2025, sending air-quality measures

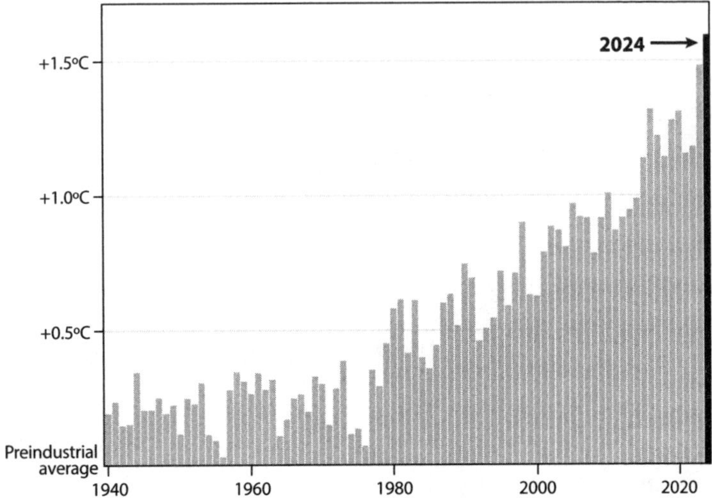

FIGURE 16.1. Global average temperatures since 1940, expressed as deviation from the average for the interval 1850–1900. At the time of this writing, 2023 and 2024 stood as the warmest recorded by meteorologists.

soaring across North America, with fire-driven haze reaching as far as Europe.) Burning in the increasingly dry Amazon rain forest removed as much forest as human activities had done in recent years. Overall, the annual loss of tree cover to fires has doubled since the opening decade of our century. Also, a sweltering heat wave settled onto the southwestern United States in July 2023, with daily temperatures in Phoenix reaching at least 110°F (43°C) for 31 straight days. And Phoenix was not alone; heat waves baked parts of India, China, Europe, and elsewhere. Sadly, in 2024 heat killed more than 1,300 pilgrims taking part in the annual hajj to Mecca.

Water warms more slowly than air, but in 2023 and again in 2024, the global ocean set its own record for mean annual temperature. Warmer waters fuel stronger storms, with places as diverse as Madagascar and Mozambique, southern California, and Libya pounded by lethal rain and wind. Unusually strong and damaging hurricanes swept over parts of the Caribbean and southeastern United States. Possible tornados touched down even in the British Midlands, not, until now, a common event.

By themselves, the recent *anni horribiles* might not seem a big deal. Weather can vary markedly from one year to the next, and I can remember shirt-sleeve weather in January when I was a student in Boston. But if we put 2023 and 2024 in context, a broad and persistent pattern emerges. It's not just that 2023 was unusually warm; global mean annual temperature has been rising for decades, and the 10 warmest years since measurements began have all occurred since 2015. Scientists no longer argue whether Earth is warming. The big question is whether global warming is accelerating. And consistent with scientific predictions for a warming planet, Earth continues to illustrate what the United Nations has called a "staggering" increase in extreme weather events.

Earth is getting hotter, and I'm guessing that most readers of this book will know why. In the twenty-first century, humans have become a major participant in the conversation between Earth and life. Humans impact the Earth system in a variety of ways, but an obvious place to begin is our influence on climate, as mediated by the carbon cycle. Each year, humans add prodigious amounts of greenhouse gases to the atmosphere, about two orders of magnitude more than all of Earth's volcanoes combined. As shown

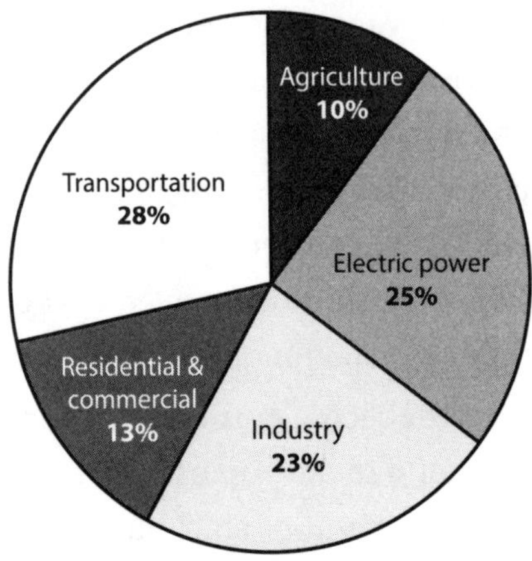

FIGURE 16.2. Sources of US greenhouse emissions in 2022.

in figure 16.2, human emissions reflect the foundational processes of modern industrial societies: electricity generation and transportation, driven mostly by fossil fuels, and industrial processes such as cement and steel manufacture, both of which generate large quantities of CO_2, at least when formed by conventional processes (more about this in a bit). Only about half of the carbon dioxide added by humans stays in the atmosphere; the rest is taken up by plants or the oceans, slowing the rate of global warming but decreasing seawater pH, which is bad news for many organisms that make skeletons of calcium carbonate.

In chapter 2, we noted that Earth has a remarkable system of temperature regulation. As the air grows warmer, rates of chemical weathering increase, removing carbon dioxide from the atmosphere and so lowering temperatures. Won't that counter human CO_2 emissions? The answer is yes, but on a timescale of hundreds or even thousands of years. Carbon dioxide added to the atmosphere this year will remain there for many generations—unless we remove it.

CO_2 is not the only contributor to our planet's increasing temperature. A bit more than 10 percent of twenty-first-century warming reflects increasing emissions of methane, or natural gas (CH_4). Methane is much less abundant in the atmosphere than carbon dioxide, but on a molecule-by-molecule basis, it is a significantly stronger contributor to greenhouse warming. About a third of the methane annually emitted to the atmosphere comes from methanogenic archaea in the digestive systems of cattle and other livestock, as well as the microbial breakdown of manure. Another 30 percent is due to the production and transport of petroleum and natural gas, much of it associated with leaks in natural gas pipelines, which recent research has shown can spew large amounts of methane into the atmosphere on short timescales.

Further influences on greenhouse warming reflect human participation in the biogeochemical cycling of nitrogen. In chapter 3, we discussed the nitrogen cycle, driven by microbial processes for billions of years. A number of bacteria and archaea fix nitrogen, converting nitrogen gas into ammonia that can be taken up and used in the formation of proteins, nucleic acids, and other biomolecules. Eukaryotic organisms do not participate in nitrogen fixation or in nitrification, denitrification, or most other metabolisms that cycle this critical element—with two known exceptions. As noted in chapter 3, *Braarudosphaera bigelowii,* a planktonic alga found in coastal waters, has established symbiotic bacteria as nitrogen-fixing organelles, much as the bacterial ancestors of mitochondria and plastids did eons ago. The second example, not surprisingly, is us. Early in the twentieth century, the German chemists Fritz Haber and Carl Bosch figured out how to produce ammonia by reacting nitrogen gas with hydrogen in the presence of an iron catalyst. An industry was born.

Today, humans fix as much nitrogen as microbes do, generating fertilizers that play a major role in feeding our planet's eight billion people. But microbes in the soil have their own uses for nitrogen fertilizer, carrying out nitrification and denitrification, which returns fixed nitrogen to the atmosphere. When these processes do not proceed to completion, they release nitrous oxide, or laughing gas (N_2O), yet another greenhouse gas. Scientists estimate that laughing gas accounts for about 6 percent of present-day greenhouse warming. Other human-generated molecules—for example, the fluorinated gases used in diverse household and commercial applications—further contribute to greenhouse warming, albeit in relatively minor ways.

The large increase in global mean temperature measured in recent years cannot easily be explained by greenhouse gases alone, prompting scientists to consider decadal variations in ocean circulation. In mid-2023, Earth transited into what is known as an El Niño state, an episodic occurrence in which changing wind patterns drive warming in surface waters of the tropical Pacific Ocean, with global consequences. El Niño certainly contributed to the magnitude of 2023 and 2024 warming but has since begun to dissipate. And yet, many climate models suggest that even in tandem, increasing greenhouse gases and El Niño do not fully explain our recent record temperatures. Perhaps, it is suggested, the additional warming reflects a decrease in cloud cover, as well as atmospheric pollutants, generally a good thing, but a change that reduces the reflection of solar radiation back into space. That places some countries in a dilemma. India, for example, has warmed much less than most other countries in recent years. Scientists continue to debate the causes of this observation, but air pollution, which causes more than a million deaths in India each year, is widely viewed as a major contributor. In similar fashion, it is predicted that Canadian wildfires will slow regional rates of warming and sea-ice melting, at the cost of decreased air quality and habitat destruction. Such considerations emphasize the complexity of understanding climate change.

So we know that Earth is growing warmer; that's recorded by many thousands of measurements taken over the past century or more. We also know from long-term measurements that current global warming parallels an increase in atmospheric greenhouse gases. And, from other, chemical analyses, we understand that humans are the primary source of these gases. As the cartoon character Pogo once intoned, "We have met the enemy and he is us."

Global warming is also evident in well-documented records of melting glaciers. Switzerland, for example, is estimated to have lost 10 percent of the water locked in its glaciers in 2022 and 2023 alone, and glaciers are retreating as well, in Greenland, the Himalaya, and Antarctica. The amount of polar sea ice also reached record lows in 2024. The addition of glacial meltwater to the oceans contributes to rising sea level, again clear from diverse and continuing observations. Because water expands as it warms, rising seawater temperatures also contribute to observed sea-level change. Measured rates of sea-level increase established yet another record for 2024: 0.23 in (0.59 cm) per year, far higher than predicted. That may not sound like much, but continued over a century, it will have serious consequences for coastal communities. In recent years, I've spent some evenings reading arguments about sea-level change by climate-change skeptics. Some make reasonable points, but with one glaring omission. For example, one recent post pointed out that in the deep past, rapid sea-level change far eclipsed that predicted for the coming century and life went on, with few coincident extinctions. True enough, but there's that omission. Put Miami, Venice, or Bangkok into the picture, and even moderate changes in sea level will be devastating. In many ways, it is the fixed geographies of cities, roads, croplands, and more that dictate the human cost of global change.

Years ago, the United Nations' Intergovernmental Panel on Climate Change argued that to avoid permanent, deleterious, environmental change, we—all of us—need to limit the increase in global temperature to no more than 1.5°C (2.7°F) above preindustrial levels. By mid-2024, that limit was breached. In fairness, as already noted, El Nino conditions pushed that year's temperatures upward, and 2025 might be a bit cooler. Nonetheless, if you welcomed a new baby in 2024, the 1.5°C threshold will be surpassed permanently (barring geoengineering fixes—see below) by the time your child enters elementary school.

A few years ago, Jean-François Bastin and colleagues at the Swiss Federal Institute of Technology published an intriguing analysis of how the climatic conditions of major world cities will change by 2050. Not surprisingly, they found that cities in the Northern Hemisphere will become warmer, making their climates similar to the current climates of cities as much as 1,000 km (620 miles) to the south. Madrid, they predict, will come to resemble present-day Marrakesh; London will become more like today's Barcelona, and Toronto will approach the current climate of Washington, D.C. Cities at lower latitudes will tend to warm less but become drier.

And, as mean temperatures increase, temperature extremes will grow more . . . well, extreme. According to Nikolaos Christidis and his colleagues at the Met Office, Britain's weather agency, daytime temperatures of 50°C (122°F)—temperatures that severely compromise basic physiological functions of the human body—were rare or impossible before the modern industrial age began. Along the Mediterranean and in the Middle East, however, they may become annual events by the end of the current century. The years 2023 and 2024 provide a glimpse of what is in store, at least if we choose to do little or nothing.

The good news is that a majority of citizens, in the United States and around the world, accept that climate change poses a serious problem for humanity. That said, in the United States, at least, 3 in 10 people see little evidence that climate change is now or will become a critical issue. Most of these people are not evil, even if they are misinformed (those who promote misinformation for financial gain are another story). Some, at least, remind me of a possibly apocryphal statement attributed to the wife of one of Charles Darwin's early critics. "My dear," she is said to have proclaimed, "descended from the apes! Let us hope it is not true, but if it is, let us pray that it will not become generally known." Head firmly in the sand, we can hope that it will all go away.

Except that it won't. In the absence of strong international effort, it will simply get worse. There aren't many climate-change deniers among those courageous individuals who fight wildfires. Nor is climate change ignored by insurance companies. In Florida, for example, nearly a dozen companies have pulled out of the home-insurance business, recognizing that stronger storms and rising sea level will increasingly affect their bottom lines. Consistent with this, an analysis published in 2024 predicts that in the United States home-insurance rates will rise rapidly over the coming decade.

Fortunately, if invention got us into this mess, invention can get us out. Hybrid automobiles, for example, sharply increase the mileage that can be achieved on a tank of gas. And electric cars are rapidly becoming more widespread. Energy-efficient heat pumps can dramatically decrease the carbon cost of heating and cooling homes and offices. New ways of manufacturing cement and steel no longer generate carbon dioxide and, in the case of cement at least, actually provide a sink for, rather than a source of, CO_2. At the same time, an ever-greater proportion of electricity is being generated by renewable sources, mainly wind, the sun, and flowing water. According to the *New York Times*, renewable sources now provide about 22 percent of all electricity in the United States, and their market share is increasing rapidly; nuclear plants supply another 18 percent (table 16.1). Indeed, both renewables and nuclear power have now eclipsed coal as a source of electricity. And novel ways of generating electricity without adding carbon dioxide to the atmosphere continue to gain attention. For example, hydrogen power—the generation of power by the oxidation of hydrogen gas to water—could emerge as an integral part of our energy future. To date, a stumbling block has been the energetic and financial cost of forming the hydrogen needed to generate electricity, but scientists and engineers have started to turn to the Earth for the required H_2. A number of crustal processes generate hydrogen, and we might be able to drill for these resources or pump water into deep, hot environments where iron-rich rocks can convert water to hydrogen for us.

Table 16.1. Percentage of power sources for electricity generation in the United States, 2000 versus 2023

Source	2000	2023
Coal	51	16
Oil	2	<1
Natural gas	17	42
Nuclear	21	18
Wind	<1	10
Solar	<1	6
Hydro	6	6

Source: Popovich, 2024.

In the twenty-first century, renewable energy is a reality, and citizens, scientists, governments, and companies all share an interest in its development. In no small part, this stems from the imperative to limit greenhouse emission, but it also reflects another aspect of humans' biogeochemical footprint: Industrial societies have dramatically increased the rate at which petroleum is removed from sedimentary rocks and oxidized to carbon dioxide, but we haven't altered the rate at which new petroleum reserves form. Petroleum forms slowly, over millions of years, so in the twenty-first century oil is a diminishing resource. With this in mind, petroleum companies and petroleum-producing countries are increasingly investing heavily in clean power. Saudi Arabians, for example, have recognized both that petroleum resources are declining and that the warming fossil fuels engender could make their country nearly uninhabitable by the close of this century. In general, global data show accelerating global investments in clean energy, a trend that is likely to increase in coming years.

Together, efficiency in energy use and innovation in energy generation promise to slow the rate of CO_2 accumulation in the atmosphere. By themselves, however, they will not decrease the amount of atmospheric greenhouse gases. For this reason, in a number of countries, plants that actively remove carbon dioxide from air are expanding, demonstrating proof of concept if not yet operating on scale. The prospect of braking global warming by actively removing greenhouse gases is real, and it shows promise, if citizens, governments, and industries commit to making it happen. There is also much discussion of a geoengineering concept in which the injection of sulfur aerosols into the stratosphere would reflect incoming solar radiation, cooling the planet, much as volcanic aerosols do. But like volcanic aerosols, aerosols injected by humans into the stratosphere would rain out in just a few years, requiring immense and continuous replenishment. Should the injection program stop, it would soon be like it had never happened, at least as far as greenhouse warming is concerned. And we still have a lot to learn about the consequences of aerosol engineering on relatively fine spatial scales. Engineered cooling could be beneficial on a global scale but disastrous in some places. (In this regard, the fact that its effects are short-lived might be seen as an advantage.) Research into geoengineering can and should continue, but investing in new sinks for atmospheric CO_2 may be our best bet for the future.

While warming temperatures and their related effects—changes in water availability, extreme weather, rising sea level, and more—demonstrate the importance of humans in the conversation between Earth and life, they are not the only ways that humans are changing the world we inhabit. Agriculture has dramatically changed the Earth's surface, with nearly two-thirds of our planet's habitable land surface now used for crops or grazing. (Grazing and the crops used to feed livestock account for almost 80 percent of agricultural lands.) Clearing natural lands for agriculture contributes to greenhouse emissions while also challenging biological diversity and increasing rates of soil erosion. It is estimated that global erosion associated with croplands has depleted the critical resource of topsoil by about half since

Victorian times. Diversion of water for irrigation likewise changes the availability of water in a number of regions around the world.

Of course, we need extensive croplands to feed the human population, so returning large areas of fields and pasture to their natural state isn't in the cards. What we can do is make more efficient use of agricultural lands. In many parts of the world, fertilizers and improved agricultural practice have dramatically increased crop yields, without which much more land would be required to feed us all. Given that the land-use cost of raising animals far exceeds that of raising crops, shifts toward more vegetable-rich diets in highly developed nations would also help greatly. (People in developing countries already have diets driven mainly by plants.)

Finally, there is all the stuff that humans make. Ron Milo, that indefatigable quantifier of nature and humans, and his colleagues have argued that in the twenty-first century, the things made by humans have come to outweigh the mass of all living organisms. That seems crazy; how can anthropogenic structures outweigh Earth's forests and reefs? To be sure, there are uncertainties in the estimates of both nature and human products, but the case is strong. Through converting natural resources to structures (especially aggregate, such as gravel and concrete), humans influence the conversation between Earth and life in yet more ways.

In recent years, a number of Earth scientists have advocated defining a new division of the geologic timescale: the Anthropocene Epoch. Agreeing on a boundary observable in the rock record has proven contentious, but the concept is clear. As anticipated by pioneers from Humboldt to Vernadsky and articulated in modern form by luminaries such as Eugene Stoermer and Paul Crutzen, the Anthropocene is the age when humans have become a dominant voice in the conversation between Earth and life. The world inhabited and in many ways fashioned by technological humans is genuinely a distinct time in our planet's history.

If you want a fascinating view of the human footprint, go to Anthroponumbers.org, a remarkable website built by Ron Milo and his partners. Here, the biological, biogeochemical, and environmental consequences of human activities are summarized in terms that are relatively easy to understand. It may be hard to grasp what it means when scientists estimate that 300 billion cubic meters of glacial ice melt annually, but the fact that this is equivalent to one shipping container of melted ice per person per year somehow makes it seem more digestible—and staggering. The annual rate of concrete production is approximately two pickup trucks per person per year. And the annual movement of sand, rocks, and soil through industry, urbanization, and agriculture amounts to some 14 pickup trucks per person. Given that the human population is about eight billion, those are big numbers. The bottom line is that twenty-first-century humans impact all of Earth's biogeochemical cycles. We are a major component of the modern Earth system.

In a previous book entitled *A Brief History of Earth*, I devoted a few pages in the last chapter to the environmental impacts of modern humans. Many readers seem to have appreciated my attempt to consider human history within the broader framework of our planet's full story, but not everyone did. I think, in particular, of a person who commented on Amazon that a generally enjoyable narrative was marred by the "doom and gloom" of the final chapter. Now, everyone is entitled to their own opinion, and scientists generally develop a reasonably thick skin, as science is, at heart, a culture of criticism. This one, however, stayed with me because I didn't set out to be gloomy. What I hoped to convey, in that book and here, was a set of facts about the world we live in and the world our grandchildren stand to inherit. Perhaps the facts are indeed gloomy, but should we not embrace the challenge before us, the future will undoubtedly be gloomier still. In a 1962 essay written for the *New York Times*, the great novelist James Baldwin made a statement that resonates strongly with twenty-first-century global change: "Not everything that is faced can be changed, but nothing can be changed until it is faced."

That said, in recent years, I've been buoyed by increasing efforts by companies, governments, scientists, engineers, and ordinary citizens to create a world of abundantly available energy that does not increase global warming, agriculture that can feed the world's people while promoting land-use practices that conserve biological diversity, and cleaner industries that lessen the still-considerable global burden of air and water pollution; in short, a world in which our grandchildren—everyone's grandchildren—can thrive. So I hope that you will see in these pages not doom and gloom, but a call to action.

CHAPTER 17

Other Conversations?

Mars

The conversation between Earth and life is both ancient and intricate. But is it unique? Are we alone in the universe, or are similar conversations distributed widely in the vastness of space? To ask this question seems fundamental to the human condition. Our ancestors considered it for thousands of years before we acquired the tools to rearticulate the question in terms of exploration. And, curiously, our generation may be among the first to entertain seriously the thought that we could be one of a kind.

Our remote ancestors were sophisticated observers of the night sky, using the positions of stars and phases of the moon as guides to time and direction. They observed that most stars move in tightly choreographed synchrony across the sky, but a few do not. Wanderers that follow their own trajectories were commonly viewed as gods or the abodes of gods. Once Copernicus, Galileo, and Kepler redrew the solar system, however, the wanderers were recognized as planets revolving, like Earth, about the Sun. Given that the wanderers were like our own planet, might they not have mountains, oceans, and life, perhaps even civilizations? Indeed, as Earth was the model, intelligence was more or less the expectation for alien life. A verse in Milton's "Paradise Lost" captures this new perspective well: "Witness this new-made World . . . with stars / Numerous, and every star perhaps a world / Of destined habitation."

In the nineteenth century, as telescopes improved, it became possible to examine Mars, our planetary neighbor, in unprecedented detail. By 1877, Nathaniel Green, a British astronomer, had produced a lovely set of maps compiled from the observations of many colleagues. Green's painterly embrace of diffuse lines and pastel shades—darker in the northern hemisphere and lighter in the south, with white caps at the poles—communicates clearly the limits of what could be seen. Within a year, however, Green's masterpiece had been eclipsed by a new map from the Italian astronomer Giovanni Schiaparelli. Where Green had artfully conveyed uncertainty, Schiaparelli drew a sharply detailed landscape dissected by broad *canali*, or channels. Schiaparelli's map provided the jumping-off point for an extraordinary fin de siècle debate about Mars, dominated by the amateur American astronomer Percival Lowell.

In 1895, Lowell published a map of Mars in which Schiaparelli's *canali* were rendered as straight lines with round structures at their intersections—canals, not channels, constructed by a parched civilization dependent for survival on meltwater from polar glaciers. The race was on, and as detailed by Jaime Green in her luminous book *The Possibility of Life*, some of the most engaging accounts of Martian life began to flow not from scientists, but, rather, from novelists. Many writers followed Lowell in painting Martians as an ancient, wise, and benevolent civilization, but, famously,

H. G. Wells imagined them as something more like humans: a vicious race bent on conquest.

The public was infatuated by Lowell's interpretations, but some scientists were more skeptical. Notably, in 1907, Alfred Russel Wallace, who half a century earlier had gained prominence for hypothesizing, independently of Darwin, that natural selection shapes evolution, published a slim volume entitled *Is Mars Habitable?* "No" would be the one-word abstract of his argument, which detailed the many reasons why animals could not flourish on the cold, dry surface of the Red Planet.

In ensuing decades, as telescopes provided ever-clearer pictures of the Martian surface, interest in Red Planet civilizations faded (although Orson Welles's famous 1938 radio broadcast of *The War of the Worlds* showed that the public's hopes and fears had not completely dissipated). As Wallace had argued, there were no canals, nor any other physical evidence of complex life on Mars. Interest in Mars as a habitable planet gained new traction, however, in 1972, when the *Mariner 9* spacecraft mapped the surface of Mars from orbit, documenting, among many other features, channel systems similar to those carved by rivers on Earth (figure 17.1). Today, Mars is cold and dry, but in the past, it was warmer and wetter, perhaps a home for microbial life, if not civilizations. In 1996, thoughts of Martian life reached renewed fever pitch with the publication of a provocative paper arguing that evidence of life was encrypted in a meteorite that originated on Mars, later to be blown into space by an energetic impact, eventually to land on Antarctic ice. The meteorite's origin was not in question, as details of its

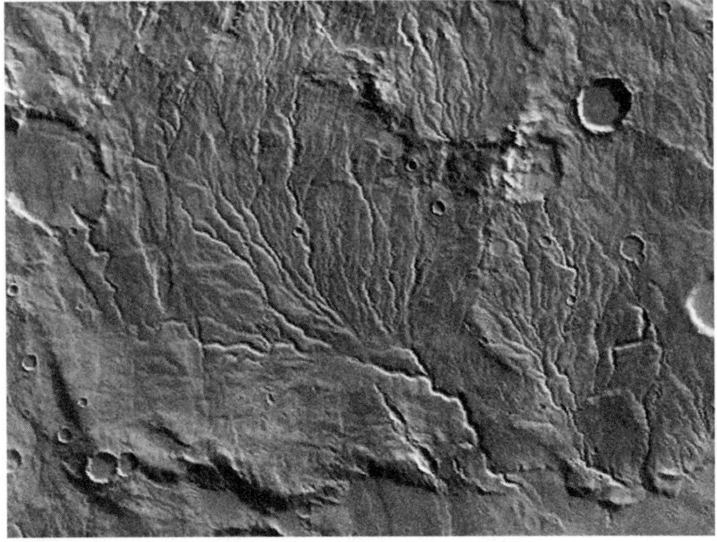

FIGURE 17.1. *Viking* orbiter image of ancient Martian drainage systems, similar to river channels observable today on Earth.

chemistry fingered Mars as its source, but interpretation of its contents proved highly contentious.

The paper's authors argued that four features observed within a meteorite called ALH-84001 collectively provide evidence of microbial Martians. First, there were tiny elongate structures interpreted as microfossils. Then, there were organic molecules called polycyclic aromatic hydrocarbons, or PAHs, as well as crystals of the iron oxide mineral magnetite, whose exceptional crystallographic purity suggested a biological origin. And finally, little blebs of carbonate minerals in cracks that traverse the meteorite were deemed similar to those induced on Earth by microbial metabolism. Fair enough, but as critics soon underscored, all of these features can be interpreted without recourse to biology. The minute "cells" are truly tiny—closer in size to the ribosomes within cells than to microbial cells per se, and their shapes are not unambiguously biological. PAHs, in turn, can be formed in many ways; they occur in the interstellar medium, among other places where biology is an unlikely option. The crystallography of the magnetites resembles that found in magnetotatic bacteria, but it has also been generated in the laboratory by purely physical processes. And the carbonate blebs really tell us only that the saturation state of fluids percolating through cracks in the meteorite's parent body reached levels high enough to drive mineral precipitation.

In time, enthusiasm for ALH-84001 as a record of Martian biology faded, but the paper had an important lasting effect: It convinced NASA that the public was deeply interested in questions of life beyond Earth, and the space agency's response helped to birth a new discipline called astrobiology. Late in 1996, the US National Research Council convened a meeting in Washington, with the goal of rearticulating NASA's planetary-exploration program in ways that gave high priority to questions of life. For NASA, the meeting inspired bold new missions to Mars and beyond. For me as well, it was transformational. On the first day of the meeting, a lanky planetary scientist named Steve Squyres introduced himself and then almost immediately popped an unexpected question: "Would you like to participate in a Mars mission?" I said, "Sure," not realizing that my research life was about to be redirected.

The mission proposal led by Steve and Ray Arvidson was duly chosen by NASA, and so, in 2004, two capable rovers,

Spirit and *Opportunity*, landed on Mars and began to move slowly across its surface, taking pictures and completing chemical analyses as they went. I was privileged to work with a remarkable team of scientists and engineers, for the first time investigating Mars much as field geologists do on Earth. *Opportunity*, the rover with which I spent most of my time, landed inside a small crater, with outcropping rocks around its margins. Almost immediately, we recognized that these were sedimentary rocks, similar to those that record the history of life and environments on Earth. *Spirit* and *Opportunity* were designed to illuminate Martian environmental history, not biology, and with careful study it soon became clear that these rocks, deposited billions of years ago, preserved both chemical and physical records of water—a critical component, of course, of habitability. The rocks themselves were sandstones, their grains consisting mostly of sulfate minerals formed by aqueous alteration of basaltic rocks that form the Martian crust. And centimeter-scale nodules within the sandstones proved to be concretions made of the iron oxide mineral hematite, precipitated from fluids that percolated through the sediments (figure 17.2; similar features can be observed on Earth). Moreover, some beds contained physical features called ripple marks, much like the sand ripples formed by water's movement along the margins of rivers and the sea. Corroborating earlier inferences from orbit, *Opportunity*'s data showed that in the past, liquid water interacted with the Meridiani surface.

In detail, the sedimentary rocks near *Opportunity*'s landing site document a regional paleoenvironment that

FIGURE 17.2. Sedimentary rocks exposed in the wall of Eagle Crater, in Meridiani Planum, Mars. Note the thin layers of sandstone and the centimeter-scale concretions, colloquially named "blueberries" for their striking hue in false color images. The concretions document oxidizing groundwater that percolated through Meridiani sediments soon after they formed. Concretions can be seen here both within the sandstone and weathered out onto the Martian surface.

was arid, acidic, and oxidizing. Evidence for aridity comes from the physical features of the sedimentary rocks, which record sand dunes dotted from time to time by sediments deposited within playa lakes. Once again, we can observe similar features on the present-day Earth—for example, in Africa's Namib Desert (figure 17.3). The fingerprint of acidity comes from a distinctive iron sulfate mineral called jarosite that forms in highly acidic waters on Earth. At a remarkable spot called Río Tinto in Spain, blood-red river waters with a pH of 1–2 precipitate jarosite and other

FIGURE 17.3. (A) Ancient sedimentary rocks exposed in the wall of Endurance Crater, Meridiani Planum, Mars. High angled layers in lowermost beds (arrow) reflect migrating sand dunes. The flat-lying beds above them also document sand migration across the surface. Not easily observed in this image, the uppermost beds found locally document playa lakes that formed transiently at this time. **(B)** Similar physical features can be seen in modern arid environments where sand dunes and playa lakes form, as shown here in the Namib Desert, Namibia.

minerals seasonally onto the riverbed. Thus, as the mission progressed, I joined a group of American and Spanish colleagues to analyze Río Tinto sediments with an instrument called a Mössbauer spectrometer, just like the one *Opportunity* used to identify iron-bearing minerals on Mars. The mineral signatures we obtained were almost identical to those in the ancient Martian rocks, cementing the view that Río Tinto provides a useful environmental analog to the ancient Meridiani record. Finally, the hematite concentrations discovered by *Opportunity* document oxic conditions; the source of the iron in those little balls was ferrous iron (Fe^{2+}) freed by aqueous alteration from basaltic rocks and then oxidized to the ferric iron (Fe^{3+}) in the hematite by at least small amounts of oxygen at the Martian surface. Despite its extreme features, Río Tinto supports a diverse microbiota, so the combination of aridity, acidity, and oxic conditions on Mars does not in and of itself preclude the presence of life. That said, *Opportunity* never observed anything that even whispered of possible microbes.

As the mission progressed, I became interested in something called water activity, because it turns out that not all water is habitable. Ions in solution form transient bonds with H_2O that keep it from doing the chemical work of water—water activity, if you will. In terms of life, water activity can be thought of in the context of salinity. Many readers will be familiar with Samuel Taylor Coleridge's poem "The Rime of the Ancient Mariner." The mariner is adrift in a small boat, hundreds of kilometers from land, surrounded by water but dying of thirst. As Coleridge memorably put it:

Water, water, every where,
And all the boards did shrink;
Water, water, every where,
Nor any drop to drink. (Wordsworth & Coleridge, 1798)

In the poem, the mariner is in distress because he shot an albatross, but biologists know that his real problem was that humans cannot long tolerate the intake of salt water. Simply put, the water activity of seawater is too low for humans.

Humans are not the only species limited by water activity. All known organisms, from bacteria to mammals, are limited in this way, although their specific tolerances vary from species to species. Most familiar organisms, including animals and plants, live in environments with a water activity of about 1.0–0.98—the water activities of pure water and seawater. A limited diversity of microbes can thrive in much saltier waters, with a few fungi and archaea able to thrive at or below 0.74, the water activity measured for a saturated solution of table salt. The record, so far as has been determined, is held by a fungus called *Xeromyces bisporus*. *Xeromyces* can live at a water activity of 0.63, although there are suggestions that the theoretical limit for life in general may be a bit lower. The reason that food preserves well in salty or sugary solutions is that their low water activities inhibit bacterial and fungal growth.

This is relevant to thinking about ancient Mars because it is possible to estimate the water activities at which different salts precipitated long ago on our neighbor. Well, to be honest, I couldn't do it, but Nick Tosca, then a postdoc in my lab and now a distinguished professor at the University of

Cambridge, could and did. The upshot is that the waters that once existed where *Opportunity* landed may have begun as relatively dilute solutions, but evaporation through time pushed them to water activities lower than any known to support life on Earth. Even with liquid water, life would have been challenging.

A second property of water that affects habitability is its persistence, and this provides further perspective on the discoveries made during the Mars Exploration Rover mission. Working on the other side of the planet, *Opportunity*'s twin, *Spirit*, made a fascinating discovery late in its mission. After supporting locomotion for more than two years, *Spirit*'s right front wheel froze and had to be dragged along as the rover continued to traverse the floor of Gusev Crater. That turned out to be a blessing in disguise, as the broken wheel churned up a white material just beneath the orangish dust that coats the crater floor. The material consisted largely of amorphous silica—that is, SiO_2, the stuff of quartz, but in a noncrystalline form. It is hard to imagine ways of forming amorphous silica in the absence of liquid water, so *Spirit* provided a new window into Mars's watery past. But as Nick Tosca and I came to realize, the amorphous silica in Gusev Crater told us something else about our neighbor's environmental history: Amorphous silica forms commonly on the surface of present-day Earth, but it rarely persists in older rocks, because continued contact with water and burial eventually transform it to crystalline quartz. Other minerals found in ancient Martian rocks by *Spirit* and *Opportunity*, including certain clays and sulfate minerals such as jarosite, similarly form in association with water but seldom persist in its continuing presence. This suggests that while water was necessary to form these minerals on the Martian surface, they have not seen much water since they formed. Mars has been cold and dry for a long time.

In telling the story of Mars, I have focused on *Spirit* and *Opportunity*, those intrepid mechanical pioneers. In part, this is because their mission signaled a remarkable new chapter in the history of solar-system exploration, but it is also because it was the one mission where I was in the room. On June 18, 2018, *Opportunity* gave up the ghost, having logged more than 5,000 Martian days and travelled more than 45 km (28 miles). (Mission success, defined by NASA before the rovers landed, was 90 Martian days of continuous operation and a traverse of 600 m, just over a third of a mile; *Oppy*'s engineers hit it out of the park. *Spirit*'s mission, which ended in 2010, was also a resounding success.)

This ended my direct participation in Mars exploration, but by the time *Opportunity* went silent a newer rover, called *Curiosity*, was hard at work in another location. And yet another rover, called *Perseverance*, would launch toward an additional target in 2020. Both *Curiosity* and *Perseverance* have capabilities beyond those of *Spirit* and *Opportunity*, and they have made, and continue to make, many important discoveries. Both have found small amounts of organic matter in sedimentary rocks deposited in ancient lakes (figure 17.4). And they've identified not only organic carbon in those lake beds but also evidence of phosphorus and nitrogen, essential elements for life on Earth. Importantly, the ancient lake beds explored by these rovers suggest the

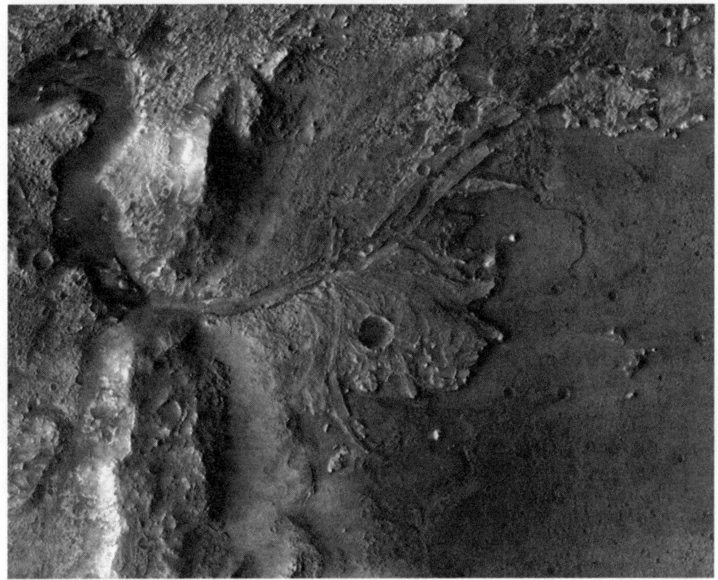

FIGURE 17.4. Ancient delta deposits preserved in Jezero Crater, site of NASA's *Perseverance* rover mission.

persistence of habitable environments over long intervals of time, perhaps tens of thousands of years. Moreover, recent evidence indicates that during at least part of those intervals, some of these lakes experienced seasonal drying, potentially facilitating the type of chemistry that led to life on Earth. (It is worth noting that NASA is not alone in probing the Red Planet. The European Space Agency's long-running [since 2003] Mars Express mission originally included both an orbiting satellite and a lander; the lander failed to deploy properly, but the mission's orbital satellite has made, and continues to make, important contributions to our understanding of our neighbor. More recently, China successfully landed a rover on Mars, although it, too, suffered malfunctions that compromised its mission. Operating a mechanical scientist on Mars is no easy feat.)

So far, like *Spirit* and *Opportunity*, *Curiosity* and *Perseverance* have not found unambiguous evidence of Martian life. For example, the organic molecules found in the ancient lake sediments could come from organic-rich micrometeorites that rained onto Mars's surface. (Earth receives more than 5,000 metric tons of these tiny meteorites each year.) But, as I was working to complete this volume, NASA reported the discovery of a thin bed of sedimentary rock in Jezero Crater that could possibly, just maybe, record the presence of microbes on the young Mars. The rock in question is a mudstone shot through with small concretions that formed within the sediments. The concretions have white centers made of calcium sulfate, with black rims of iron phosphate minerals. On Earth, such things can form by localized microbial activities within accumulating sediments, a fact that occasioned some breathless internet articles on the discovery. NASA spokespeople have been clear in pointing out that nothing observed so far *requires* the participation of organisms, although it cannot be ruled out, and for a number of scientists, life is the preferred alternative. Samples have been cached for future return to Earth, and they should tell us whether life was involved in the formation of this rock. But be patient, as Mars sample return remains some years into the future.

Whatever the eventual interpretation of the unusual Jezero rock, *Curiosity* and *Perseverance* have sharply increased our understanding of Mars's environmental history through their careful studies of lake beds and, in *Percy*'s case, an ancient delta. And, in light of their discoveries, questions of persistence again raise important issues. As already noted, habitable water bodies may have been sustained for 10,000 years or more at these locations. Ten thousand years is a long time, but the implications for biology differ depending on whether early Mars was more or less continuously habitable for many millions of years or only episodically habitable, with temperate conditions existing for, perhaps, 10,000 or 100,000 out of every 10 million years. To me, at least, evidence increasingly points toward the latter view.

As happens so often in science, conversation over afternoon coffee resulted in a collaboration with my colleague Robin Wordsworth. Robin, a gifted atmospheric scientist, generated a model of ancient Mars in which a generally cold and dry planetary surface was episodically turned wet and warm by greenhouse gases produced by large meteorite impacts and/or massive volcanism. After each warming event, Mars slowly returned to its cold and dry state as the greenhouse gases dissipated. Through time, as both the number and size of impacting meteorites and the frequency and volume of volcanic eruptions declined, the probability that potentially habitable conditions would recur sank to exceedingly low values. We aren't the only ones advocating for episodic water in Mars's early history, although other mechanisms have been proposed. For now, however, debate over the timing and persistence of warm, wet environments on early Mars persists, and further insights will be welcome.

Since *Spirit* and *Opportunity* landed in 2004, our understanding of Mars as a planet has increased tremendously.

Many questions remain, and it is certainly too early to declare that Mars never gave rise to life. Indeed, some people speculate that Mars might still harbor microorganisms, sustained by chemical reactions in wet environments deep within the crust. Testing this presents formidable challenges, and it is unclear whether subterranean life can long persist in the absence of continuing subsidies from a surface biota. But I can't prove it wrong.

In general, increasing observations have produced little that might document the kinds of conversation between planet and life that have transformed Earth through time, tempering astrobiological expectation. Most scientists would agree that if life ever existed at the Martian surface, it disappeared long ago. Present-day Mars has no plate tectonics; its atmosphere has largely dissipated through time; and liquid water at the surface is rare and transient. This reemphasizes a point made earlier: Perhaps what is unusual about Earth is not that life began there, but that it has persisted for some four billion years. In the prologue to his great novel *Red Mars*, Kim Stanley Robinson declared, "We are all the consciousness that Mars has ever had." We don't know yet, but just possibly, on that eventual day when astronauts set foot on Mars, they will be the first life of any kind to grace our neighbor.

CHAPTER 18

Other Conversations?

Everywhere Else

Given its promising physical features and proximity, Mars was the obvious target for humanity's first efforts in "boots on the ground" astrobiological exploration, but it is not our solar system's only body of interest. Venus, today, is something of an anti-Mars, its mean surface temperature a scorching 464°C (867°F), maintained by a runaway greenhouse atmosphere. Some, however, propose that earlier in its history, Venus was more temperate, perhaps a potential abode for life. It has even been suggested that the clouds that envelop the present-day Venusian surface are sufficiently cool to support life. Indeed, phosphine (PH_3), a gas produced by organisms on Earth, has been reported from Venusian clouds, although both the measurement and its interpretation have been debated vigorously. Neither do all planetary scientists agree that the surface of Venus was once habitable. Thus, pending new space missions, some in the planning stage, we can say little more about ancient Venusian habitability. For now, Earth remains the inner solar system's Goldilocks planet for life: Mars is too cold, and Venus too hot. Earth? Just right.

The concept of a habitable zone—the range of orbits around a star within which liquid water can be stable on revolving planets—permeates modern discussions of astrobiology. Actually, however, the idea goes back a long way. In his revolutionary *Philosophiæ Naturalis Principia Mathematica* (1687), Isaac Newton estimated the range of distances from the Sun within which planets could sustain liquid water. Of course, we now know that the surface temperature of a planet reflects the greenhouse composition of its atmosphere as well its parent star's luminosity and proximity. Many scientists now refer to this potential zone of life as the circumstellar habitable zone, because, as it turns out, there is more than one way to sustain liquid water in our solar system.

Europa is a moon of Jupiter, far too distant from the Sun to be heated by its rays. Despite this, Europa has an ocean hidden beneath a veneer of water ice. Beginning in the 1970s, analysis of light absorbed or reflected by Europa identified H_2O on its surface. Later satellite images confirmed the moon's icy face, and ensuing measurements of Europa's gravity demonstrated that this surface is only skin deep, extending downward 80–170 km (50–106 miles) to a rocky interior. Finally, studies of magnetism indicated that the lower part of Europa's watery mantle is liquid. Together, then, light, gravity, and magnetism revealed a subsurface ocean deep within the solar system.

How can liquid water be maintained so far from the Sun? The answer is "tides." Most readers, whether they live in Atlantic City or Saskatoon, are familiar with tides on Earth.

Our planet and moon are locked in a gravitational dance, and as Earth rotates beneath the moon, seawater is alternately drawn toward the moon or, on the other side of our planet, away from it, generating the oscillating tides observed along coastlines. (The Sun also influences tides on Earth, much as the moon does, but less strongly.) Tides affect the solid Earth as well, but because both Earth and the moon are minor players in the gravitational relationships found throughout our solar system, and because the moon's orbit is nearly circular, tidal influence on the solid Earth is small. Not so for the moons revolving around our solar system's giant planets, Jupiter and Saturn. Together, these planets have some 369 documented moons, mostly small bodies with highly eccentric orbits. (No fewer than 128 of Saturn's are sufficiently tiny that they were discovered only in early 2025.) Jupiter's strong gravitational pull induces tides in the moons, and as their rocky interiors are pushed and pulled, the resulting friction generates heat. Io, the closest moon to Jupiter, gets so hot that its rocky interior melts; because of this, Io is our solar system's most active volcanic body. The next three closest moons, Europa, Ganymede, and Callisto, don't generate volcanoes, but they heat up enough to melt the lower part of their icy surfaces.

Currently, much astrobiological interest focuses on Europa. Light cannot penetrate beneath the moon's icy surface, but geochemical models suggest that chemical reactions between Europa's ocean and its rocky interior could provide energy for at least a limited biosphere. Indeed, magnetic data indicate that Europa's subsurface ocean is salty, telling us that that water does interact chemically with

FIGURE 18.1. Europa, an astrobiologically interesting moon of Jupiter, imaged by NASA's *Galileo* spacecraft. Cracks in Europa's icy surface enable communication between the moon's surface and the ocean below.

underlying rocks. Moreover, the strong tides induced by Jupiter crack Europa's icy shell, allowing subsurface ocean water to spread onto the surface, depositing sodium chloride (NaCl, or table salt) and perhaps other materials on top of the ice (figure 18.1). There is also evidence for carbon dioxide ice (the "dry ice" of high school science demonstrations), documenting the presence of carbon at and near Europa's surface. Thus, observations of Europa check some of the boxes of interest to astrobiologists. Liquid water? Check. Source of energy? Check. Carbon? Check. But is there nitrogen? Phosphorus? We don't yet know.

For the moment, we can't tell whether Europa's subsurface ocean is habitable, perhaps even inhabited. If all goes well, however, we'll soon know much more. In October 2024, NASA launched its *Europa Clipper*, a mission that promises to enhance dramatically our understanding of the Europan ocean. Importantly, a suite of instruments will interrogate the chemistry of Europa's surface, providing a deeper understanding of the chemical makeup of subsurface ocean waters that episodically reach the surface. Those fissures in Europa's ice shell may also transport surface materials, including organic matter from micrometeorites, downward into the ocean. Direct observations of microorganisms are unlikely, but the mission team will be alert for possible indications of a conversation between moon and life in the details of Europa's surface chemistry. The *Clipper* should arrive at Europa in 2030.

Jupiter is not alone in harboring moons of astrobiological interest. Indeed, two of the most fascinating moons in the solar system revolve around Saturn: Enceladus and Titan.

Enceladus is a small moon, less than 600 km (375 miles) in diameter, with an icy surface that sits atop a rocky interior (figure 18.2a). So far, this sounds a bit like Europa, but there are intriguing differences. Notably, Enceladus's north polar region is pocked by numerous impact craters, evidence that this surface has existed for a long time, while the southern hemisphere is nearly smooth, revealing that it has been resurfaced relatively recently. A series of long, broadly parallel fissures colloquially known as tiger stripes marks the south polar region. Observations by NASA's *Cassini* spacecraft have revealed that the temperature of these stripes is high relative to the rest of the moon's surface.

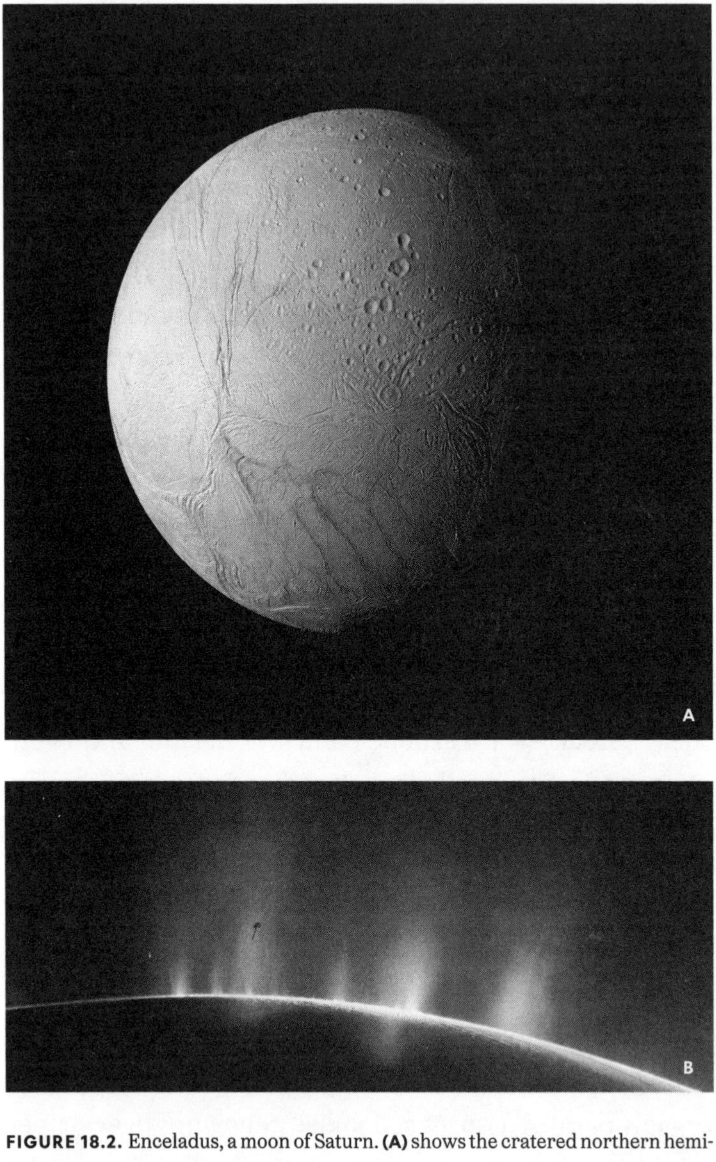

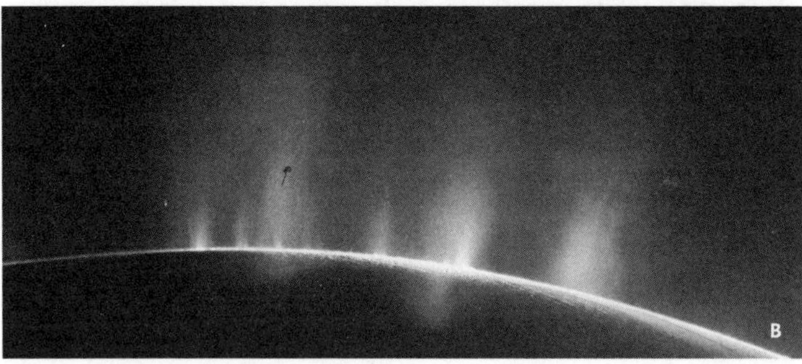

FIGURE 18.2. Enceladus, a moon of Saturn. **(A)** shows the cratered northern hemisphere and the "tiger stripes" that mark the southern hemisphere, from which **(B)** geyser-like fountains spew.

And *Cassini* discovered something else: The stripes put on one of the solar system's greatest shows. Geyser-like fountains regularly spew subsurface liquids hundreds of kilometers into space (figure 18.2b). Like Europa, Enceladus has a subsurface ocean, and it is the spectacular emissions from this ocean that continually resurface its southern hemisphere.

As was true for Europa, we will not soon drill through Enceladus's ice cover to sample its ocean directly. But once again, we can gain some understanding of that ocean from the materials delivered to the surface by the fountains. And Enceladus offers an additional option not available for Europa. We can sample the jets of salty water as they shoot into space. Indeed, this has been done, and future missions will include new-wave instruments that will enable more sophisticated chemical analyses. So far, we know that Enceladus's subsurface ocean is salty, indicating, once again, that subsurface waters react chemically with the moon's rocky interior. We also know that there is hydrogen (H_2), nitrogen gas, traces of ammonia (NH_3), and even amines—carbon-bearing molecules bound to hydrogen- and nitrogen-rich structures, found, among other places, in the amino acids that form proteins. Other carbon-bearing compounds have also been detected, including carbon dioxide and trace amounts of simple organic molecules, such as methane, propane, acetylene, formaldehyde, and benzene. None of these molecules requires the presence of life, but a number of them—formaldehyde, for example, and those amines—are thought to have played a role in the origin of life on Earth. Proposed missions are designed to detect the possible presence of true biomolecules and perhaps even microbial cells in Enceladus's magnificent fountains. We don't know what future missions will find. Enceladus may or may not harbor a limited microbial biota. At the very least, however, Enceladus will be confirmed as another solar-system body where biologically relevant carbon chemistry takes place.

This leaves Titan, perhaps the most interesting body in the solar system, save for Earth itself. Titan is the largest moon of Saturn. It has a thick, cloudy atmosphere and a surface patterned by lakes and rivers. This sounds a bit like Earth, and the first images of Titan's surface, beamed home by the European Space Agency's *Huygens* probe that dropped onto the moon in 2005, look eerily familiar (figure 18.3). That familiarity, however, is misleading. Titan's atmosphere consists mostly of nitrogen gas, but its clouds are hydrocarbons; it rains methane, and its rivers and lakes consist of liquid methane as well. The surface of Titan is far too cold for liquid water; the mean surface temperature is −179°C (−290°F). There is H_2O at the surface of Titan, however; it makes up the moon's solid surface. There may be liquid water at depth, heated, once again, by tidal friction. I doubt that we'll ever detect life on Titan, but its remarkable chemistry can tell us much about the possibilities of environments and carbon chemistry far from Earth.

FIGURE 18.3. Boulders of water ice on the surface of Titan, imaged by the *Huygens* probe.

In sum, we may or may not be the only life to have taken root in our solar system. Mars and, perhaps, Venus might have harbored organisms in the past, and, just possibly, microbes may exist today in the subsurface oceans of moons that orbit Jupiter and Saturn. I'm not convinced that any of these possibilities is likely, but their probability is not zero, and continuing exploration is warranted. What we can say with more certainty is that within our solar system, only on Earth has the conversation between life and its physical home transformed both through time. And only on Earth has persistent life evolved intelligent beings capable of asking questions about the universe we inhabit.

From our earthbound perspective, the solar system seems immense, but on the scale of the universe as a whole, it is no more than a tiny speck in the vastness of space. Our parent galaxy, the Milky Way, contains more than 100 billion stars, and it is estimated that 100–200 billion galaxies (or more!) are distributed throughout the universe. That's a lot of stars, and most of them are thought to have one or more planets. If only one planet (or moon) in a billion sustains life, the universe must be a lively place. For this reason, many scientists, including me, doubt that we are alone in the universe. A 2024 survey conducted by Peter Vickers and colleagues found that 85 percent or more of scientists believe strongly or moderately that life exists elsewhere in the cosmos, and this holds whether the scientists in question are physicists or biologists, astrobiologists or not. A smaller but still substantial percentage also think that intelligent life is out there somewhere. The challenge is how to test such conjectures.

When I began my scientific career, the only planets known to science were those in our own solar system. By the early 1990s, however, astronomers had begun to detect planets in orbit around other stars. And by August 14, 2025, 5,983 extrasolar planets had been confirmed, a number that continues to grow rapidly. The discovery of planetary abundance prompts two questions: How do we detect planets in faraway solar systems, and how can we ask questions about extrasolar life?

As planets revolve around their parent star, gravity influences both bodies, causing the star to wobble just a bit. In consequence, the wavelengths of light received from the star vary regularly, their "beat" determined by the period of the revolving planet. To date, more than a thousand planets have been detected in this way. A relatively small number of planets have also been detected by other methods involving the influence of the star's and planet's gravitational influence on light. A few have even been observed directly. By far the largest number of discoveries, however, have been made by a method called "transit."

To understand the transit method, think first of eclipses, like the total eclipse of the Sun seen by many North Americans in 2024. In these events, the moon passes between Earth and the Sun, transiently blocking solar radiation. The moon is much smaller than the Sun, but because it is much nearer to us, its disk pretty much obscures the Sun during a total eclipse. In similar fashion, a large number of extrasolar planets have orbits that, with each revolution, position them transiently in the line of sight between their parent star and Earth. When the planet passes in front of it,

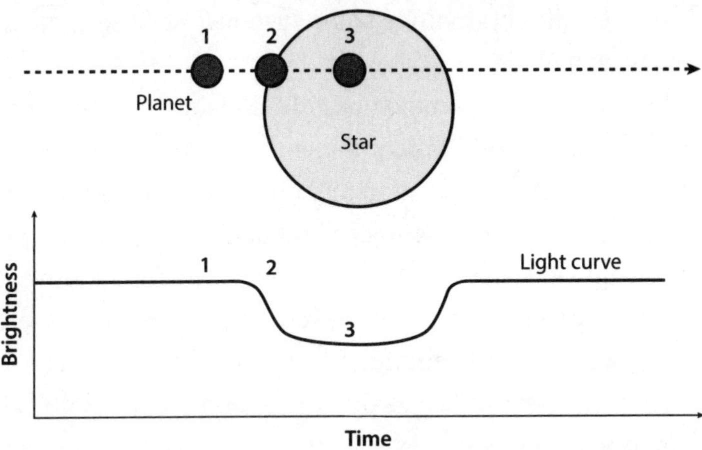

FIGURE 18.4. The transit method of extrasolar planet detection. As each orbit of the planet places it transiently in front of its star, the light from the star is dimmed.

the light received from the star is dimmed—once per orbit (figure 18.4). The amount of dimming is tiny, because the size of orbiting planets is much smaller than the stars they orbit, but it can be quantified from precise measurements. Such data enable astronomers to determine not only the presence of a planet but also its diameter. Combined with mass calculations based on gravitational wobble, this permits determination of the planet's density. Is it a rocky body like Earth, a gas giant like Jupiter, or something in between, perhaps rich in liquid water or ice? Also, if the planet can be observed long enough, the regular dimming permits calculation of its period, the time it takes to complete one orbit. Not surprisingly, early research detected mostly giant planets in tight orbits around their star. With improved instruments and longer observations, however, it has become clear that rocky planets broadly like Earth occur commonly in other solar systems, including those closest to our own.

Recently, it has become possible to make one more critical measurement. Some planets have atmospheres, and by measuring the reflection and absorption of light intercepted by these atmospheres and comparing it to unimpeded light from its star, we can gain some understanding of the atmosphere's composition. Recent measurements have even revealed the three-dimensional pattern of circulation in the atmosphere of a nearby exoplanet. Thus, in combination, the planet's internal and atmospheric compositions, along with its distance from its parent star and the star's luminosity, allow inferences about potential habitability. A brave new world, indeed.

How can we assess the possibilities of life on extrasolar planets? One thing is certain: We can't go there; even the closest extrasolar planets are much too far away for orbiters and rovers. To put this in perspective, NASA launched two satellites in 1977, bound for the farthest reaches of the solar system. After a diversion to investigate Jupiter, Saturn, and their moons, *Voyager 1* reached the outer edge of the solar system in 2012; *Voyager 2* followed in 2018. Both continue on their outward voyages, the first human-built structures to reach interstellar space. At 4.2 light-years' distance, Proxima Centauri is the closest star to our solar system. Were one or both *Voyagers* to deadhead it toward this star, it would take them well over a thousand years to get there. And even if we could deploy an intelligent rover on one of Proxima Centauri's planets, it would take more than four years for a single instruction to reach our mechanical scientist, and four more to receive its reply. (With Mars, it takes about eight minutes.)

Absent the types of orbiters and landers that are providing a new understanding of our own solar system, there are two routes forward. First, we can ask which of the extrasolar planets discovered to date lie within their star's habitable zone. The news here is relatively good. Earth may be the Goldilocks planet in our solar system, but on the scale of the universe, it is hardly unique. Of course, we also need to consider the stars that keep extrasolar planets in their gravitational thrall. The planet Proxima Centauri b, for example, lies within Proxima Centauri's habitable zone as conventionally defined, but because Proxima Centauri is what astronomers call an M-dwarf, it pummels its planets with ultraviolet radiation in doses strong enough to jeopardize any atmosphere.

The ability to interpret the atmospheres of extrasolar planets suggests a second, more specific, way to listen for a conversation between planet and life. James Lovelock, introduced early in this book as a pioneer in thinking about Earth and life in integrated fashion, once stated flatly that anyone could see from afar that Earth is a biological planet while Mars is not. His reasoning was that in Earth's atmosphere oxidizing gases like oxygen exist alongside reducing gases such as methane. In the absence of life, oxygen and methane would interact, quantitatively removing one of them. Only life, with its continuing fluxes of both oxygen and methane, could keep our atmosphere far from chemical equilibrium over long intervals of time.

Many years ago, at a dinner, I tried Lovelock's logic out on David Stevenson, one of my generation's preeminent planetary scientists. After listening to my recital, David smiled slowly and then said, "I can guarantee you that when that measurement is made, within six months there will be half a dozen models of how to achieve this by physical processes alone." Maybe, or maybe not, but the broader point is well taken. Scientists continue to debate whether atmospheres can provide diagnostic evidence of planet (or moon) and life in conversation. (A coda: Tim Lyons and his colleagues have suggested that, even if confirmed as a biosignature, methane-oxygen disequilibrium would not have been detectable to a distant viewer of Earth over its entire history, yielding a false negative for alien astrobiologists.)

Ever more powerful telescopes, epitomized, for the moment, by the extraordinary James Webb Space Telescope, are now imaging the heavens in unprecedented detail. Recently, for example, JWST found a rocky extrasolar planet cloaked by an atmosphere. Sound familiar? Continuing observations have also detected extrasolar planets at increasingly large distances from Earth. At present, the most distant planet thought to be potentially habitable is Kepler-1606 b, some 2,870 light-years away. More distant planets are known—currently the limit is about 27,000 light-years—but their immense distances make it difficult to make statements about habitability. Given that the most distant objects in the observable universe are some 47 billion light-years away, it is clear that most of space remains beyond our ability to make astrobiological observations.

The only way we'll learn about life in the greater universe is if extraterrestrials introduce themselves. With that in mind, the SETI Institute in California has long been listening for evidence of intelligent beings. Despite the terrific 1997 movie *Contact*, we've yet to receive unambiguous messages from space. In fairness, probing the entire universe for signals from distant technologies is a long process, and much of the cosmos remains to be assayed. Nonetheless, the silence to date prompts a big question, said to have been asked in casual conversation by the Nobel Prize–winning physicist Enrico Fermi: "Where is everybody?"

The Fermi Paradox, named after Fermi in recognition of his deceptively simple query, focuses on the discrepancy, for now at least, between the expectation that space should be richly endowed with life, including intelligent life, and the lack of evidence for such a universe. Perhaps our expectations are wrong, and we are indeed the only instance of technological intelligence in the universe. Or perhaps others are out there but have not sent messages our way. Novelists and filmmakers have commonly imagined contact between earthlings and faraway civilizations. Perhaps, as H. G. Wells imagined, extraterrestrials are malevolent, bent on conquest. In contrast, *The Day the Earth Stood Still*, perhaps the Cold War era's greatest science fiction film, depicts a peaceful ambassador from space who delivers a stern message to humans: Forsake your violent ways or suffer annihilation.

Other options are possible. Perhaps aliens, again sensing humans' warlike ways, simply don't want us to know about them. Or perhaps they could care less. Fermi's own speculation was that technologically advanced civilizations eventually destroy themselves. At present, commentary on the Fermi Paradox runs more to philosophy than science, but keeping a sophisticated ear on space is our one way to probe the question.

What, then, does our accelerating exploration of the universe say about us? About Earth? Whether Earth is unique in sustaining life—and, as noted earlier, I suspect that we are not—our planet may well be special not only in providing the environmental conditions for life to originate but also in sustaining and supporting life through some four billion years of planetary change. Earth's transformation through time owes much to physical processes, none more important than plate tectonics. These processes, originating in the mantle, have given rise to stable continents, with their mountains and plains, as well as majestic arrays of volcanoes, many emerging from the sea. Life, in turn, has made possible the environmental transitions that converted an anoxic planet into one whose oxygen-rich surface can support complex multicellular organisms. Importantly, the world in which we live is not simply a function of physical Earth history; nor has it been shaped exclusively by organisms. Our Earth—our home—reflects interactions between physical and biological processes that have played out over billions of years: the conversation between Earth and life. In an age when humans have become a major voice in this conversation, there is an urgent need to understand the integrated Earth system and our role in it much better than we do at present. Our future depends on it.

ACKNOWLEDGMENTS

Curiosity, it is said, killed the cat. Perhaps it did, but for scientists, curiosity is catnip. It is why we get up in the morning and why we endure the long hours and repeated failures that line the road to discovery. I was born curious, and the older I get, the more I appreciate the many small ways in which my parents supported my curiosity. My sainted Aunt Kay did as well, sensing in the youthful me possibilities I couldn't imagine, again and again handing me books designed to broaden my horizons.

Throughout my student years, dedicated teachers stoked my innate curiosity, and when I got to graduate school, I really hit the jackpot. My thesis advisor, Elso Barghoorn, took me into the field and challenged me to think hard about both Earth and life. The other members of my thesis committee—Dick Holland, Steve Gould, Bernie Kummel, and Ray Siever—almost daily encouraged me to think in new (for me) ways. And I was fortunate to befriend Boston University's Steve Golubic, then and now a world expert on cyanobacteria. And in a summer at the Woods Hole Oceanographic Institution before graduate school, Dave Johnson patiently taught me how to do research. All of these wonderful teachers contributed in important ways to my development as a scientist.

Since those early days, I've been fortunate to work with a diversity of colleagues in biology and Earth sciences, each, consciously or unconsciously, helping me to look beyond disciplinary boundaries in my attempts to understand the Earth and its history. Our planet's history is written in stone, and I am eternally grateful for the many scientists with whom I've shared both the pleasures and rigors of fieldwork. Brian Harland invited me to join his team in Spitsbergen in 1979, an opportunity that changed my research life. Keene Swett and I teamed up a few years later for a partnership that took us to Spitsbergen, Greenland, and the Canadian Arctic over the next decade. From there, the world really opened up, enabling me to learn about the geology (and culture) of Russia, including Siberia, from Misha Fedonkin, Petr Kolosov, Misha Semikhatov, and Volodya Sergeev; China from Zhang Yun; Australia from Malcolm Walter; and southern Africa from Gerard Germs, Nic Beukes, and John Grotzinger. Beyond fieldwork, long-term collaborations with John Grotzinger, Dick Bambach, John Hayes, and Pupa Gilbert have done much to shape my worldview. Students have as well, and I know that I've been extremely fortunate to work with a series of exceptional undergraduates, graduate students, and postdoctoral fellows during my years at Harvard. I thank them all.

Tim Lyons, who graciously read my draft manuscript, suggested that the book should be published as a memoir. It isn't really a memoir, at least not by design, but I haven't shied away from using my own experiences to illuminate many of the topics explored in these pages. The challenges and rewards of my own research—ranging from detailed work on the nature of early life and environments, plant and animal radiations, and mass extinctions to Mars exploration—have done a great deal to shape my thinking about the conversation between Earth and life.

In a possibly apocryphal anecdote, it is said that one year, as Albert Einstein handed out an exam to his graduate class at Princeton, his student assistant, concern evident on his face, noted that it was the same test that Einstein had given the previous year. "Yes," replied Einstein, "it is the same test, but the answers have changed." In this spirit, I note that some of the discussions in this volume revisit topics I've addressed previously in earlier books. My intent has not been to replicate earlier arguments but to update them and place them in a somewhat different framework.

I thank Jim Kasting and Tim Lyons for their careful reading of my first draft. Their support and, especially, their constructive criticisms are greatly appreciated. I am also indebted to the team at Princeton University Press, especially my editor Alison Kalett. Alison's guidance and thoughtful edits did much to improve this book. Also, I acknowledge that several passages in this volume are reworked from reviews I've written for the (London) *Times Literary Supplement*, especially the historical discussions in chapters 1 and 17.

Lastly, I thank my family. Without the abiding love and support of my beloved wife Marsha, my career would have been very different; in her absence, I doubt that you would be reading this book. I am also grateful to my children, Kirsten and Rob, who, among many other things, continually remind me that some things in life are more important than science.

REFERENCES AND FURTHER READING

For readers who wish to dig deeper, here are some references to major points covered in each chapter. There are both classic and recent books, technical publications, and a few websites that provide sound information in a digestible format. Asterisks mark particularly accessible works. Many of the scientific papers cited can be accessed through Google Scholar.

CHAPTER 1

Barghoorn, E. S., and Tyler, S. A. (1965). Microorganisms from the Gunflint chert. *Science, 147,* 563–577.

Canfield, D. E. (Ed.). (2016). *Baas Becking's geobiology.* Wiley-Blackwell.

Cloud, P. E. (1968). Atmospheric and hydrospheric evolution on the primitive Earth. *Science, 160,* 729–736.

Darwin, C. (1859). *On the origin of species by means of natural selection* (1st ed.). J. Murray.

Helferich, G. (2004). *Humboldt's cosmos: Alexander von Humboldt and the Latin American journey that changed the way we see the world.* Gotham.

Holland, H. D. (2006). The oxygenation of the atmosphere and oceans. *Philosophical Transactions of the Royal Society B, 361,* 903–915.

Hutton, J. (1788). Theory of the Earth; or an investigation of the laws observable in the composition, dissolution, and restoration of land upon the globe. *Transactions of the Royal Society of Edinburgh, 1,* 209–304.

Knoll, A. H., Canfield, D. E., and Konhauser, K. (Eds.). (2012). *Fundamentals of geobiology.* Wiley-Blackwell.

Konhauser, K. O. (2006). *Introduction to geomicrobiology.* Wiley-Blackwell.

*Lovelock, J. (1979). *Gaia: A new look at life on Earth.* Oxford University Press.

Lovelock, J. E. (1989). Geophysiology. *Transactions of the Royal Society of Edinburgh: Earth Sciences, 80*, 169–175.

Lyell, C. (1990). *Principles of geology* (Vols. 1–3). University of Chicago Press. (Original work published 1830–1833)

*Perman, R. (2023). *James Hutton: The genius of time*. Birlinn.

Rudwick, M.J.S. (1997). *Georges Cuvier, fossil bones, and geological catastrophes: New translations and interpretations of the primary texts*. University of Chicago Press.

*Rudwick, M.J.S. (2014). *Earth's deep history: How it was discovered and why it matters* (Illustrated ed.). University of Chicago Press.

Vernadsky, V. (1996). *The biosphere* (English ed.). Springer. (Original work published 1926)

*Wulf, A. (2015). *The invention of nature: Alexander von Humboldt's new world*. Knopf.

CHAPTER 2

Aiuppa, A., and others. (2019). CO_2 flux emissions from the Earth's most actively degassing volcanoes, 2005–2015. *Scientific Reports, 9*, Article 5442. https://doi.org/10.1038/s41598-019-41901-y

Allègre, C. J., Poiter, J.-P., Humler, E., and Hofmann, A. (1995). The chemical composition of the Earth. *Earth and Planetary Science Letters, 134*, 515–526.

*Bar-On, Y. M., Phillps, R., and Milo, R. (2018). The biomass distribution on Earth. *Proceedings of the National Academy of Sciences, 115*, 6506–6511.

*Bralower, T., and Bice, D. (n.d.). *Overview of the carbon cycle from a systems perspective*. Pennsylvania State University. https://www.e-education.psu.edu/earth103/node/1019

Dixon, R. K., and others. (1994). Carbon pools and flux of global forest ecosystems. *Science, 263*, 185–190.

*EarthHow. (n.d.). *What is the carbon cycle?* https://earthhow.com/carbon-cycle/

Falkowski, P. G. (2012). The global carbon cycle: Biological processes. In A. H. Knoll, D. E. Canfield, and K. O. Konhauser (Eds.), *Fundamentals of geobiology* (pp. 5–19). Wiley-Blackwell.

Fischer, R. A., and others. (2020). The carbon content of Earth and its core. *Proceedings of the National Academy of Sciences, 117*, 8743–8749.

Gottschalk, G. (1979). *Bacterial metabolism*. Springer.

*Hazen, R. M. (2019). *Symphony in C: Carbon and the evolution of (almost) everything* (Illustrated ed.). W. W. Norton & Company.

Hazen, R. M., Jones, A. P., and Baross, J. A. (Eds.). (2013). Carbon in Earth. *Reviews in Mineralogy and Geochemistry, 75*, 1–698.

*Helmenstine, A. M. (2018, June 28). *The elemental composition of the human body*. ThoughtCo. https://www.thoughtco.com/elemental-composition-of-human-body-603896

Malucelli, M., and others. (2016). Where is it and how much? Mapping and quantifying elements in single cells. *Analyst*, *141*, 5221–5235.

Marty, B., and Tolstikhin, I. N. (1998). CO_2 fluxes from mid-ocean ridges, arcs and plumes. *Chemical Geology*, *145*, 233–248.

Morrison, S. M., and others. (2020). Exploring carbon mineral systems: Recent advances in C mineral evolution, mineral ecology, and network analysis. *Frontiers in Earth Science*, *8*. https://doi.org/10.3389/feart.2020.00208

*Nealson, K. H. (1997). Sediment bacteria: Who's there, what are they doing, and what's new? *Annual Review of Earth and Planetary Sciences*, *25*, 403–434.

Orcutt, B., Dasgupta, R., and Daniel, I. (Eds.). (2019). *Deep carbon: Past to present*. Cambridge University Press.

Plank, T., and Langmuir, C. H. (1998). The chemical composition of subducting sediment and its consequences for the crust and mantle. *Chemical Geology*, *145*, 325–394.

Wallman, K., and Aloisi, G. (2012). The global carbon cycle: Geological processes. In A. H. Knoll, D. E. Canfield, and K. O. Konhauser (Eds.), *Fundamentals of geobiology* (pp. 20–35). Wiley-Blackwell.

CHAPTER 3

Barton, A. D., Dutkiewicz, S., Flierl, G., Bragg, J., and Follows, M. J. (2010). Patterns of diversity in marine phytoplankton. *Science*, *327*, 1509–1511.

Canfield, D. R., and Farquhar, J. (2012). The global nitrogen cycle. In A. H. Knoll, D. E. Canfield, and K. O. Konhauser (Eds.), *Fundamentals of geobiology* (pp. 49–64). Wiley-Blackwell.

Ciampitti, A. I., and others. (2021). Revisiting biological nitrogen fixation dynamics in soybeans. *Frontiers in Plant Science*, *12*. https://doi.org/10.3389/fpls.2021.727021

Coale, T. H., and others. (2024). Nitrogen-fixing organelle in a marine alga. *Science*, *384*, 217–222.

Elser, J. J., and others. (2007). Global analysis of nitrogen and phosphorus limitation of primary producers in freshwater, marine and terrestrial ecosystems. *Ecology Letters*, *10*, 1135–1142.

Filippelli, G. M. (2002). The global phosphorus cycle. *Reviews in Mineralogy and Geochemistry*, *48*, 391–425.

*Filippelli, G. M. (2008). The global phosphorus cycle: Past, present, and future. *Elements*, *4*, 89–95.

Föllmi, K. B. (1996). The phosphorus cycle, phosphogenesis and marine phosphate-rich deposits. *Earth-Science Reviews, 40,* 55–124.

Gruber, N., and Galloway, J. N. (2008). An Earth-system perspective of the global nitrogen cycle. *Nature, 451,* 293–296.

Kowalczewska-Madura, K., and Goldyn, R. (2012). Spatial and seasonal variability of pore water phosphorus concentration in shallow Lake Swarzędzkie, Poland. *Environmental Monitoring and Assessment, 184,* 1509–1516.

Kwon, E. Y., and others. (2022). Nutrient uptake plasticity in phytoplankton sustains future ocean net primary production. *Science Advances, 8,* Article eadd2475.

Le Bauer, D. S., and Treseder, K. K. (2008). Nitrogen limitation of net primary productivity in terrestrial ecosystems is globally distributed. *Ecology, 89,* 371–379.

Martin, J. H., and Fitzwater, S. E. (1988). Iron deficiency limits phytoplankton growth in the north-east Pacific subarctic. *Nature, 331,* 341–343.

Menzel, D. W., and Ryther, J. H. (1960). The annual cycle of primary production in the Sargasso Sea off Bermuda. *Deep-Sea Research, 6,* 351–367.

*Moore, C. M., and others. (2013). Processes and patterns of oceanic nutrient limitation. *Nature Geoscience, 6,* 701–710.

Palya, A. P., and others. (2011). Storage and mobility of nitrogen in the continental crust: Evidence from partially melted metasedimentary rocks, Mt. Stafford, Australia. *Chemical Geology, 281,* 211–226.

Paytan, A., and McLaughlin, K. (2007). The oceanic phosphorus cycle. *Chemical Reviews, 107,* 563–576.

Postgate, J. R. (1982). *The fundamentals of nitrogen fixation.* Cambridge University Press.

*Redfield, A. C. (1958). The biological control of chemical factors in the environment. *American Scientist, 46,* 205–221.

*Sterner, R. W., and Elser, J. J. (2002). *Ecological stoichiometry: The biology of elements from molecules to the biosphere.* Princeton University Press.

Stüeken, E. E., Pellerin, A., Thomazo, C., Johnson, B. W., Duncanson, S., and Schoepfer, S. D. (2024). Marine biogeochemical nitrogen cycling through Earth's history. *Nature Reviews Earth & Environment, 5,* 732–747.

Ward, B. (2012). The global nitrogen cycle. In A. H. Knoll, D. E. Canfield, and K. O. Konhauser (Eds.), *Fundamentals of geobiology* (pp. 36–48). Wiley-Blackwell.

CHAPTER 4

Charlson, R. J., Lovelock, J. E., Andreae, M. O., and Warren, S. G. (1987). Oceanic phytoplankton, atmospheric sulphur, cloud albedo and climate. *Nature, 326,* 655–661.

Ehrlich, H. L., Newman, D. K., and Kappler, A. (2015). *Ehrlich's geomicrobiology* (6th ed.). CRC Press.

Farooq, M. A., and Dietz, K.-J. (2015). Silicon as versatile player in plant and human biology: Overlooked and poorly understood. *Frontiers in Plant Science, 6,* Article 994.

Fike, D. A., and others. (2015). Rethinking the ancient sulfur cycle. *Annual Review of Earth and Planetary Sciences, 43,* 593–622.

Gruber, N., and Deutsch, C. A. (2014). Redfield's evolving legacy. *Nature Geoscience, 7,* 853–855.

Gutjahr, C. (2025). Outsmarted by fungi. *Science, 387,* Article 927928.

Jørgensen, B. B., and others. (2019). The biogeochemical sulfur cycle of marine sediments. *Frontiers in Microbiology, 10,* Article 849. https://doi.org/10.3389/fmicb.2019.00849

Joyce, G., and Szostak, J. (2018). Protocells and RNA self-replication. *Cold Spring Harbor Perspectives in Biology.* https://doi.org/10.1101/cshperspect.a034801

Kappler, A., and Straub, K. L. (2005). Geomicrobiological cycling of iron. *Reviews in Mineralogy and Geochemistry, 59,* 85–108.

Kendall, B., and others. (2012). The global iron cycle. In A. H. Knoll, D. E. Canfield, and K. O. Konhauser (Eds.), *Fundamentals of geobiology* (pp. 65–92). Wiley-Blackwell.

Lovelock, J. E., Maggs, R. J., and Rasmussen, R. A. (1972). Atmospheric dimethyl sulphide and the natural sulphur cycle. *Nature, 237,* 452–453.

Marron, A. O., and others. (2016). The evolution of silicon transport in eukaryotes. *Molecular Biology and Evolution, 33,* 3226–3248.

Melton, E. D., and others. (2014). The interplay of microbially mediated and abiotic reactions in the biogeochemical Fe cycle. *Nature Reviews Microbiology, 12,* 797–808.

Moore, C. M., and others. (2013). Processes and patterns of oceanic nutrient limitation. *Nature Geoscience, 6,* 701–710.

Morel, F.M.M., and Price, N. M. (2003). The biogeochemical cycles of trace metals in the oceans. *Science, 300,* 944–947.

Sutak, R., and others. (2020). Iron uptake mechanisms in marine phytoplankton. *Frontiers in Microbiology, 11,* Article 566691.

Tagliabue, A., and others. (2017). The integral role of iron in ocean biogeochemistry. *Nature, 54,* 51–59.

Thiel, E. C. (2004). Iron, ferritin and nutrition. *Annual Review of Nutrition, 24,* 327–343.

Tsang, T., and others. (2021). Copper biology. *Current Biology, 31,* R415–R429.

Widdel, F., and others. (1993). Ferrous iron oxidation by anoxygenic phototrophic bacteria. *Nature, 362,* 834–836.

Williams, R.J.P., and Fraústo da Silva, J.J.R. (1997). *The natural selection of the chemical elements: The environment and life's chemistry*. Clarendon Press.

CHAPTER 5

*Ball, P. (2000). *H_2O: A biography of water*. Weidenfeld & Nicolson.

Boccaletti, G. (2021). *Water: A biography*. Penguin Random House.

Broecker, W. (2010). *The great ocean conveyor: Discovering the trigger for abrupt climate change*. Princeton University Press.

*Canfield, D. E. (2014). *Oxygen: A four billion year history*. Princeton University Press.

Fabian, P., and Dameris, M. (2014). *Ozone in the atmosphere: Basic principles, natural and human impacts*. Springer Verlag.

Kasting, J. F., and Canfield, D. E. (2012). The global oxygen cycle. In A. H. Knoll, D. E. Canfield, and K. O. Konhauser (Eds.), *Fundamentals of geobiology* (pp. 93–104). Wiley-Blackwell.

Keeling, R. F. (1995). The atmospheric oxygen cycle: The oxygen isotopes of atmospheric CO_2 and O_2 and the O_2/N_2 ratio. *Reviews of Geophysics, 33*, 1253–1262.

*Lane, N. (2020). *Oxygen: The molecule that made the world* (Rev. ed.). Oxford University Press.

*National Oceanic and Atmospheric Administration. (2023, March 24). The hydrologic cycle. https://www.noaa.gov/jetstream/atmosphere/hydro

Peslier, A. (2020). The origins of water. *Science, 369*, 1058.

Ruben, S., Randall, M., Kamen, M., and Hyde, J. L. (1941). Heavy oxygen (O^{18}) as a tracer in the study of photosynthesis. *Journal of the American Chemical Society, 63*, 877–879.

*Venditto, B., and LePan, N. (2021, December 14). *Visualizing the abundance of elements in the Earth's crust*. World Economic Forum. https://www.weforum.org/stories/2021/12/abundance-elements-earth-crust/

von Caemmerer, S., and Baker, N. (2007). The biology of transpiration. From guard cells to globe [Introduction to special issue]. *Plant Physiology, 143*. https://doi.org/10.1104/pp.104.900213

*Water Science School. (2019, October 25). *The distribution of water on, in, and above the Earth*. United States Geological Survey. https://www.usgs.gov/media/images/distribution-water-and-above-earth

CHAPTER 6

Adam, P. S., and others. (2017). The growing tree of Archaea: New perspectives on their diversity, evolution and ecology. *The IMSE Journal, 11*, 2407–2425.

Bar-On, Y. M., Phillps, R., and Milo, R. (2018). See chapter 2.

Baum, D. A., and Smith, S. D. (2013). *Tree thinking: An introduction to phylogenetic biology*. Roberts and Company.

*Brainard, J., and Henderson, R. (2025). *7.2 Primate classification and evolution*. FlexBooks 2.0, CK-12 College Human Biology. https://flexbooks.ck12.org/cbook/ck-12-college-human-biology-flexbook-2.0/section/7.2/primary/lesson/primate-classification-and-evolution-chumbio/

Coleman, G. A., and others. (2021). A rooted phylogeny resolves early bacterial evolution. *Science*, *372*, eabe0511.

Darwin, C. (1859). See chapter 1.

Eme, L. (2017). Archaea and the origin of eukaryotes. *Nature Reviews in Microbiology*, *15*, 711–723.

Eme, L., and others. (2023). Inference and reconstruction of the heimdallarchaeial ancestry of Eukaryotes. *Nature*, *618*, 992–999.

Gest, H. (2004). The discovery of microorganisms by Robert Hooke and Antoni van Leeuwenhoek, fellows of the Royal Society. *Notes and Records: The Royal Society Journal of the History of Science*, *58*, 187–201.

Gray, M. W. (2017). Lynn Margulis and the endosymbiont hypothesis: 50 years later. *Molecular Biology of the Cell*, *28*, 1285–1287.

Katscher, F. (2004). The history of the terms prokaryotes and eukaryotes. *Protist*, *155*, 257–263.

*Knoll, A. H. (2003). *Life on a young planet: The first three billion years of evolution on Earth*. Princeton University Press.

*Margulis, L. (1971). Symbiosis and evolution. *Scientific American*, *225*, 48–61.

*Martin, W., and Mentel, M. (2010). The origin of mitochondria. *Nature Education*, *3*(9), 58.

Martin, W. F., and others. (2016). Physiology, phylogeny, and LUCA. *Microbial Cell*, *3*, 582–587.

*Morell, V. (1997). Microbiology's scarred revolutionary. *Science*, *276*, 699–702. https://doi.org/10.1126/science.276.5313.699

Pace, N. R. (2009). Mapping the tree of life: Progress and prospects. *Microbiology and Molecular Biology Reviews*, *73*, 565–576.

Pozzi, L., and others. (2014). Primate phylogenetic relationships and divergence dates inferred from complete mitochondrial genomes. *Molecular Phylogenetics and Evolution*, *75*, 165–183.

Saap, J. (2005). The prokaryote-eukaryote dichotomy: Meanings and mythology. *Microbiology and Molecular Biology Reviews*, *69*, 292–330.

Spang, A., and others. (2015). Complex archaea that bridge the gap between prokaryotes and Eukaryotes. *Nature*, *521*, 173–179.

Stanier, R. Y., and van Niel, C. B. (1962). The concept of a bacterium. *Archiv für Mikrobiologie*, *42*, 17–35.

*Sulloway, F. J. (2005). The evolution of Charles Darwin. *Smithsonian Magazine*, *36*, 58–69.

Woese, C. R., and Fox, G. E. (1977). Phylogenetic structure of the prokaryotic domain: The primary kingdoms. *Proceedings of the National Academy of Sciences*, *74*, 5088–5090.

*Yong, E. (2017, July 17). The man who blew the door off the microbial world. *The Atlantic*. https://www.theatlantic.com/science/archive/2017/07/the-man-who-blew-the-door-off-the-microbial-world/534246/

CHAPTER 7

Bindeman, I., and others. (2018). Rapid emergence of subaerial landmasses and onset of a modern hydrologic cycle 2.5 billion years ago. *Nature*, *557*, 545–548.

*Blakey, R. (n.d.). *Deep time maps*. https://deeptimemaps.com

Brenner, A. R., and others. (2020). Paleomagnetic evidence for modern-like plate motion velocities at 3.2 Ga. *Science Advances*, *6*. https://doi.org/10.1126/sciadv.aaz8670

Cawood, P. A., and others. (2022). Secular evolution of continents and the Earth system. *Reviews of Geophysics*, *60*, Article e2022RG000789.

Dong, J., and others. (2021). Constraining the volume of Earth's early oceans with a temperature-dependent mantle water storage capacity model. *AGU Advances*, *2*, Article e2020AV000323.

du Toit, A. L. (1937). *Our wandering continents: An hypothesis of continental drifting*. Oliver & Boyd.

*Grotzinger, J. P., and Jordan, T. (2019). *Understanding Earth* (8th ed.). W. H. Freeman.

Harrison, T. M. (2009). The Hadean crust: Evidence from >4 Ga zircon. *Annual Review of Earth and Planetary Sciences*, *37*, 479–505.

Hess, H. (1962). The history of ocean basins. In A.E.J. Engel, H. L. James, and B. F. Leonard (Eds.), *Petrologic studies: A volume in honor of A. F. Buddington* (pp. 599–620). The Geological Society of America.

*Langmuir, C. H., and Broecker, W. (2012). *How to build a habitable planet: The story of Earth from the big bang to humankind*. Princeton University Press.

Oreskes, N. (Ed.). (2003). *Plate tectonics: An insider's history of the modern theory of the Earth* [eBook]. CRC Press. (Original work published 2003)

Spencer, C. J., and others. (2021). Enigmatic mid-Proterozoic orogens: Hot, thin, and low. *Geophysical Research Letters*, *48*, Article e2021GL093312.

*Tucker, N. (2021, August 9). *Marie Tharp: Mapping the ocean floor*. Library of Congress Blogs. https://blogs.loc.gov/loc/2021/08/marie-tharp-mapping-the-ocean-floor/

*United States Geological Survey. (2014). *Understanding plate motions*. https://pubs.usgs.gov/gip/dynamic/understanding.html

Wegener, A. (2011). *The origin of continents and oceans* (English ed.). Dover Press. (Original work published 1915)

CHAPTER 8

Allwood, A. C., and others. (2009). Controls on development and diversity of Early Archean stromatolites. *Proceedings of the National Academy of Sciences*, *106*, 9548–9555.

Benner, S. A., and others. (1989). Modern metabolism as a palimpsest of the RNA world. *Proceedings of the National Academy of Sciences*, *86*, 7054–7058.

Catling, D. C., and Zahnle, K. J. (2020). The Archean atmosphere. *Science Advances*, *6*, eaax1420.

Darwin, C. (1871). *Letter to Joseph Hooker*. Darwin Correspondence Project, Letter 7471. http://www.darwinproject.ac.uk/DCP-LETT-7471

*Deamer, D. (2019). *Assembling life: How can life begin on Earth and other habitable planets?* Oxford University Press.

Fry, I. (2000). *The emergence of life on Earth: A historical and scientific overview*. Rutgers University Press.

*Fry, I. (2006). The origins of research into the origins of life. *Endeavour*, *30*, 24–28.

Goldman, A. D., and Kacar, B. (2021). Cofactors are remnants of life's origin and early evolution. *Journal of Molecular Evolution*, *89*, 127–133.

Homann, M. (2018). Earliest life on Earth: Evidence from the Barberton Greenstone Belt, South Africa. *Earth-Science Reviews*, *196*. https://doi.org/10.1016/j.earscirev.2019.102888

Javaux, E. (2019). Challenges in evidencing the earliest traces of life. *Nature*, *572*, 451–460.

Joyce, G., and Szostak, J. (2018). Protocells and RNA self-replication. *Cold Spring Harbor Perspectives in Biology*. https://doi.org/10.1101/cshperspect.a034801

Kirsebom, L. A., Liu, F., and McClain, W. H. (2004). The discovery of a catalytic RNA within RNase P and its legacy. *Journal of Biological Chemistry*, *300*, Article 107318.

Miller, S. L. (1953). A production of amino acids under possible primitive Earth conditions. *Science*, *117*, 528–529.

Peretó, J., and others. (2009). Charles Darwin and the origin of life. *Origins of Life and Evolution of Biospheres*, *39*, 395.

Powner, M., and Sutherland, J. (2011). Prebiotic chemistry: A new modus operandi. *Philosophical Transactions of the Royal Society B*, *366*, 2870–2877.

*Sasselov, D. D., and others. (2022). The origin of life as a planetary phenomenon. *Science Advances*, *6*, eaax3419.

Sleep, N. (2018). Geological and geochemical constraints on the origin and evolution of life. *Astrobiology*, *18*, 1199–1219.

Smith, E., and Morowitz, H. (2016). *The origin and nature of life on Earth: The emergence of the fourth geosphere*. Cambridge University Press.

Weiss, M. C., and others. (2016). The physiology and habitat of the last universal common ancestor. *Nature Microbiology*, *1*, 1–8.

White, H. B. (1976). Coenzymes as fossils of an earlier metabolic state. *Journal of Molecular Evolution*, *7*, 101–104.

CHAPTER 9

Anbar, A., and others. (2007). A whiff of oxygen before the Great Oxidation Event? *Science*, *317*, 1903–1906.

Bekker, A., and others. (2010). Iron formation: The sedimentary product of a complex interplay among mantle, tectonic, oceanic, and biospheric processes. *Economic Geology*, *105*, 467–508.

Canfield, D. E. (2014). See chapter 5.

Farquhar, J., and Wing, B. A. (2003). Multiple sulfur isotopes and the evolution of the atmosphere. *Earth and Planetary Science Letters*, *213*, 1–13.

Fischer, W. W., and others. (2016). Evolution of oxygenic photosynthesis. *Annual Review of Earth and Planetary Sciences*, *44*, 647–683.

Flament, N., and others. (2008). A case for late-Archaean continental emergence from thermal evolution models and hypsometry. *Earth and Planetary Science Letters*, *275*, 326–336.

Hao, J., and others. (2020). Cycling phosphorus on the Archean Earth: Part II. Phosphorus limitation on primary production Archean oceans. *Geochimica et Cosmochimica Acta*, *280*, 360–377.

Holland, H. D. (2006). The oxygenation of the atmosphere and oceans. *Philosophical Transactions of the Royal Society B*, *361*, 903–915.

Johnson, A. C., and others. (2021). Reconciling evidence of oxidative weathering and atmospheric anoxia on Archean Earth. *Science Advances*, *7*, eabj0108.

Jones, C., and others. (2015). Iron oxides, divalent cations, silica, and the early Earth phosphorus crisis. *Geology*, *43*, 135–138.

Olejarz, J., Iwasa, Y., Knoll, A. H., and Nowak, M. (2021). The Great Oxygenation Event as a consequence of ecological dynamics modulated by planetary change. *Nature Communications*, *12*. https://doi.org/10.1038/s41467-021-23286-7

Ostrander, C., and others. (2024). Onset of coupled atmosphere–ocean oxygenation 2.3 billion years ago. *Nature*. https://doi.org/10.1038/s41586-024-07551-5

*Poulton, S., and Canfield, D. E. (2011). Ferruginous conditions: A dominant feature of the ocean through Earth's history. *Elements*, *7*, 107–112.

Rasmussen, B., Muhling, J. R., Tosca, N. J., and Fischer, W. W. (2023). Did nutrient-rich oceans fuel Earth's oxygenation? *Geology*, *51*, 444–448.

*Sanchez Baracaldo, P., and others. (2022). Cyanobacteria and biogeochemical cycles through Earth's history. *Trends in Microbiology*, *30*, 143–157.

CHAPTER 10

Anbar, A. D., and Knoll, A. H. (2002). Proterozoic ocean chemistry and evolution: A bioorganic bridge? *Science*, *297*, 1137–1142.

Brocks, J. J. (2023). Lost world of complex life and the late rise of the eukaryotic crown. *Nature*, *618*, 767–773.

Burki, F., and others. (2020). The new tree of eukaryotes. *Trends in Ecology and Evolution*, *35*, 43–55.

Canfield, D. E. (1998). A new model for Proterozoic ocean chemistry. *Nature*, *396*, 450–453.

Gilleaudeau, G. J., and others. (2019). Uranium isotope evidence for limited euxinia in mid-Proterozoic oceans. *Earth and Planetary Science Letters*, *521*, 150–157.

Javaux, E. J., and Knoll, A. H. (2017). Micropaleontology of the Lower Mesoproterozoic Roper Group, Australia. *Journal of Paleontology*, *91*, 199–229.

Liu, X., and others. (2021). A persistently low level of atmospheric oxygen in Earth's middle age. *Nature Communications*, *12*, Article 351.

*Lyons, T., and others. (2014). The rise of oxygen in Earth's early ocean and atmosphere. *Nature*, *506*, 307–315.

Mills, D. B., and others. (2024). Constraining the oxygen requirements for modern microbial eukaryote diversity. *Proceedings of the National Academy of Sciences*, *121*, Article e2303754120.

Moody, E.R.R., and others. (2024). The nature of the last universal common ancestor and its impact on the early Earth system. *Nature Ecology and Evolution*. https://doi.org/10.1038/s41559-024-02461-1

Riedman, L. A., and others. (2023). Early eukaryotic microfossils of the late Palaeoproterozoic Limbunya Group, Birrindudu Basin, northern Australia. *Papers in Palaeontology*, *9*(6), e1538.

Santana-Molina, C., Williams, T. A., Snel, B., and Spang, A. (2025). Chimeric origins and dynamic evolution of central carbon metabolism in eukaryotes. *Nature Ecology and Evolution*. https://doi.org/10.1038/s41559-025-02648-0

Shen, Y., and others. (2002). The chemistry of mid-Proterozoic oceans: Evidence from the McArthur Basin, northern Australia. *American Journal of Science, 302*, 81–109.

Shen, Y., and others. (2003). Evidence for low sulphate and deep water anoxia in a mid-Proterozoic marine basin. *Nature, 423*, 632–635.

Vosseberg, J., and others. (2021). Timing the origin of eukaryotic cellular complexity with ancient duplications. *Nature Ecology and Evolution, 5*, 92–100.

CHAPTER 11

Bobrovskiy, I., Nagovitsyn, A., Hope, J. M., Luzhnaya, E., and Brocks, J. J. (2022). Guts, gut contents, and feeding strategies of Ediacaran animals. *Current Biology, 32*, 5382–5389.

Cohen, P. A., and Riedman, L. A. (2015). It's a protist-eat-protist world: Recalcitrance, predation, and evolution in the Tonian–Cryogenian ocean. *Emerging Topics in Life Sciences, 2*, 173–180.

Dayle, M. J., and others. (2011). Cell differentiation and morphogenesis in the colony-forming choanoflagellate *Salpingoeca rosetta*. *Developmental Biology, 357*, 73–82.

Droser, M. L., and others. (2017). The rise of animals in a changing environment: Global ecological innovation in the late Ediacaran. *Annual Review of Earth and Planetary Sciences, 45*, 593–617.

Erwin, D., and Valentine, J. (2013). *The Cambrian explosion: The construction of animal biodiversity*. W. H. Freeman.

Erwin, D., and others. (2011). The Cambrian conundrum: Early divergence and later ecological success in the early history of animals. *Science, 334*, 1091–1097.

Grazhdankin, D. (2014). Patterns of evolution of the Ediacaran soft-bodied biota. *Journal of Paleontology, 88*, 269–283.

Hoshino, Y. (2017). Cryogenian evolution of stigmasteroid biosynthesis. *Science Advances, 3*, Article e1700887.

*Knoll, A. H. (2017). Food for early animal evolution. *Nature, 528*, 528–530.

Knoll, A. H., and Lahr, D.J.G. (2016). Fossils, feeding and the evolution of complex multicellularity. In K. Niklas and S. Neumann (Eds.), *The origins and consequences of multicellularity* (pp. 3–16). MIT Press.

Laakso, T. A., and others. (2020). Ediacaran reorganization of the marine phosphorus cycle. *Proceedings of the National Academy of Sciences, 117*, 11961–11967.

Mills, D. B., and others. (2014). Oxygen requirements of the earliest animals. *Proceedings of the National Academy of Sciences, 111*(11), 4168–4172.

Narbonne, G. M., and others. (2009). Reconstructing a lost world: Ediacaran rangeomorphs from Spaniard's Bay, Newfoundland. *Journal of Paleontology, 83*, 503–523.

Nursall, J. R. (1959). Oxygen as a prerequisite to the origin of the metazoa. *Nature, 183*, 1170–1172.

Sperling, E. A., and Vinther, J. A. (2010). Placozoan affinity for *Dickinsonia* and the evolution of late Proterozoic metazoan feeding modes. *Evolution and Development, 12*, 201–209.

Sperling, E. A., and others. (2013). Oxygen, ecology, and the Cambrian radiation of animals. *Proceedings of the National Academy of Sciences, 110*, 13446–13451.

Squire, R. J., and others. (2006). Did the Transgondwanan Supermountain trigger the explosive radiation of animals on Earth? *Earth and Planetary Science Letters, 250*, 116–133.

Stockey, R. G., and others. (2024). Sustained increases in atmospheric oxygen and marine productivity in the Neoproterozoic and Palaeozoic Eras. *Nature Geoscience*. https://doi.org/10.1038/s41561-024-01479-1

Tarhan, L. G., Droser, M. L., Planavsky, N. J., and Johnston, D. T. (2015). Protracted development of bioturbation through the early Palaeozoic Era. *Nature Geoscience, 8*, 865–869.

Tong, K., Bozdag, G. O., and Ratcliff, W. C. (2022). Selective drivers of simple multicellularity. *Current Opinion in Microbiology, 67*, Article 102141.

*Xiao, S., and Laflamme, M. (2008). On the eve of animal radiation: Phylogeny, ecology and evolution of the Ediacara biota. *Trends in Ecology and Evolution, 24*, 31–40.

Zhu, Z., Campbell, I. H., Allen, C. M., Brock, J. J., and Chen, B. (2022). The temporal distribution of Earth's supermountains and their potential link to the rise of atmospheric oxygen and biological evolution. *Earth and Planetary Science Letters, 580*, Article 117391.

CHAPTER 12

Addadi, L., and others. (2006). Mollusk shell formation: A source of new concepts for understanding biomineralization processes. *Chemistry: A European Journal, 12*, 980–987.

Benzerara, K. (2011). Significance, mechanisms and environmental implications of microbial biomineralization. *Comptes Rendus Géoscience, 343*(2–3), 160–167.

Canfield, D. A., and Raiswell, R. (1991). Carbonate precipitation and dissolution. *Topics in Geobiology, 1991*, 411–453.

Cantine, M., and others. (2020). Carbonates before skeletons: A database approach. *Earth Science Reviews*, *201*, Article 103065.

Cohen, P. A., and Knoll, A. H. (2012). Neoproterozoic scale microfossils from the Fifteen Mile Group, Yukon Territory. *Journal of Paleontology*, *86*, 775–800.

Cohen, P. A., Strauss, J. V., Rooney, A. D., Sharma, M., and Tosca, N. (2017). Controlled hydroxyapatite biomineralization in an ~810 million-year-old unicellular eukaryote. *Science Advances*, *3*, Article e1700095.

Conley, D. J., and Carey, J. C. (2015). Silica cycling over geologic time. *Nature Geoscience*, *8*, 431–432.

De Yoreo, J. J., and others. (2015). Crystallization by particle attachment in synthetic, biogenic, and geologic environments. *Science*, *349*, aaa6760.

*Dove, P. (2010). The rise of skeletal biominerals. *Elements*, *6*, 37–42.

Dove, P., de Yoreo, J., and Weiner, S. (Eds.). (2003). Biomineralization. *Reviews in Mineralogy and Geochemistry*, *54*, 1–386.

Gilbert, P.U.P.A., and others. (2022). Biomineralization: Integrating mechanism and evolutionary history. *Science Advances*, *8*, eabl9653.

Grant, S. W. (1990). Shell structure and distribution of *Cloudina*, a potential index fossil for the terminal Proterozoic. *American Journal of Science*, *290*, 261–294.

*Knoll, A. H. (2003). Biomineralization and evolutionary history. In P. Dove, J. de Yoreo, and S. Weiner (Eds.), *Biomineralization* (pp. 329–356). *Reviews in Mineralogy and Geochemistry*, *54*.

Knoll, A. H., and Follows, M. J. (2016). A bottom-up perspective on ecosystem change in Mesozoic oceans. *Proceedings of the Royal Society B: Biological Sciences*, *283*, Article 20161755.

Lefèvre, C. T., and Bazylinski, D. A. (2013). Ecology, diversity, and evolution of magnetotactic bacteria. *Microbiology and Molecular Biology Reviews*, *77*, 497–526.

Maldonado, M., Carmona, M. G., Uriz, M. J., and Cruzado, A. (1999). Decline in Mesozoic reef-building sponges explained by silicon limitation. *Nature*, *401*, 785–788.

Maliva, R., Knoll, A. H., and Siever, R. (1989). Secular change in chert distribution: A reflection of evolving biological participation in the silica cycle. *Palaios*, *4*, 519–532.

Mann, S. (2001). *Biomineralization: Principles and concepts in bioinorganic materials chemistry*. Oxford University Press.

Pasteris, J. (2008). Bone and tooth mineralization: Why apatite? *Elements*, *4*, 97–104.

Petrucciani, A., and others. (2022). Si decline and diatom evolution: Insights from physiological experiments. *Frontiers in Marine Science*, *9*, Article 924452.

Porter, S. (2010). Calcite and aragonite seas and the de novo acquisition of carbonate skeletons. *Geobiology*, *8*, 256–277.

Pruss, S., and others. (2010). Carbonates in skeleton-poor seas: New insights from Cambrian and Ordovician strata of Laurentia. *Palaios*, *25*, 73–84.

Raven, J. A., and Knoll, A. H. (2010). Non-skeletal biomineralization by eukaryotes: Matters of moment and gravity. *Geomicrobiology Journal*, *27*, 572–584.

CHAPTER 13

*Beerling, D. (2017). *The emerald planet: How plants changed Earth's history*. Oxford University Press.

Berner, R. A. (1997). The rise of plants and their effect on weathering and atmospheric CO_2. *Science*, *276*, 544–546.

Boyce, C. K., and Nelsen, M. P. (2025). Terrestrialization: Toward a shared framework for ecosystem evolution. *Paleobiology*. https://doi.org/10.1017/pab.2024.15

Buatois, L. A., and others. (2022). The invasion of the land in deep time: Integrating Paleozoic records of paleobiology, ichnology, sedimentology, and geomorphology. *Integrative and Comparative Biology*, *62*, 297–331.

Case, N. T., and others. (2025). Fungal impacts on Earth's ecosystems. *Nature*, *638*, 49–57.

Dahl, T. W., and others. (2010). Devonian rise in atmospheric oxygen correlated to the radiations of terrestrial plants and large predatory fish. *Proceedings of the National Academy of Sciences*, *107*, 17911–17915.

Edmond, J. M., and others. (1995). The fluvial geochemistry and denudation rate of the Guayana Shield in Venezuela, Colombia, and Brazil. *Geochimica et Cosmochimica Acta*, *59*, 3301–3325.

*Evans, M. (2021). *Soil: The incredible story of what keeps the Earth, and us, healthy*. Murdoch Books.

Gensel, P. G. (2012). The earliest land plants. *Annual Review of Ecology, Evolution, and Systematics*, *39*, 459–477.

Gibling, M. R., and Davies, N. S. (2012). Palaeozoic landscapes shaped by plant evolution. *Nature Geoscience*, *5*, 99–105.

Ielpi, A., and Lapôtre, M.G.A. (2020). A tenfold slowdown in river meander migration driven by plant life. *Nature Geoscience*, *13*(1), 82–86.

*Kenrick, P. (2020). *A history of plants in fifty fossils*. Smithsonian Books.

Matthaeus, W., and others. (2023). A systems approach to understanding how plants transformed Earth's environment in deep time. *Annual Reviews in Earth and Planetary Sciences*, *51*, 551–580.

Montanez, I. P., and others. (2016). Climate, pCO_2 and terrestrial carbon cycle linkages during late Palaeozoic glacial-interglacial cycles. *Nature Geoscience*, *9*, 824–828.

*Shubin, N. (2008). *Your inner fish: A journey into the 3.5-billion-year history of the human body*. Pantheon Books.

Stolper, D. A., and Keller, C. B. (2018). A record of deep-ocean dissolved O_2 from the oxidation state of iron in submarine basalts. *Nature, 553*, 323–327.

*Strullu-Derrien, C., and others. (2019). The Rhynie chert. *Current Biology, 29*, R1218–23.

Strullu-Derrien, C., and others. (2023). Hapalosiphonacean cyanobacteria (Nostocales) thrived amid emerging embryophytes in an Early Devonian (407-million-year-old) landscape. *iScience, 27*, Article 107338.

*Taiz, L., and others. (2022). *Plant physiology and development* (7th ed.). Sinauer.

*Vermeij, G. (2017). How the land became the locus of major evolutionary innovations. *Current Biology, 27*, 3178–3182.

Wolfe, J. M., and others. (2024). Convergent adaptation of true crabs (Decapoda: Brachyura) to a gradient of terrestrial environments. *Systematic Biology, 73*, 247–262.

CHAPTER 14

Boyce, C. K., and Lee, J. E. (2011). An exceptional role for flowering plant physiology in the expansion of tropical rainforests and biodiversity. *Proceedings of the Royal Society B, 277*, 3437–3443.

Brantley, S. L., and others. (2023). How temperature-dependent silicate weathering acts as Earth's geological thermostat. *Science, 379*, 382–389.

Cenozoic CO_2 Proxy Integration Project Consortium. (2023). Toward a Cenozoic history of atmospheric CO_2. *Science, 382*, eadi5177.

*Earle, S. (2023). *A brief history of the Earth's climate: Everyone's guide to the science of climate change.* New Society Publishers.

Eldridge, D. J., and others. (2012). Animal foraging as a mechanism for sediment movement and soil nutrient development: Evidence from the semi-arid Australian woodlands and the Chihuahuan Desert. *Geomorphology, 157–158*, 131–141.

Evans, D. A., Beukes, N. J., and Kirschvink, J. L. (1997). Low-latitude glaciation in the Palaeoproterozoic Era. *Nature, 386L*, 262–266.

Feulner, G. (2012). The faint young sun problem. *Reviews of Geophysics, 50*. https://doi.org/10.1029/2011RG000375

*Frontiers for Young Minds. (2023, November 7). *What is albedo and what does it have to do with global warming?* https://kids.frontiersin.org/articles/10.3389/frym.2023.1113553

Harland, W. B. (1964). Critical evidence for a great infra-Cambrian glaciation. *Geologische Rundschau, 54*, 45–61.

*Hoffman, P., and others. (n.d.). *Snowball Earth*. https://www.snowballearth.org

Hoffman, P. F., Arnaud, E., Halverson, G. P., and Shields-Zhou, G. (2011). A history of Neoproterozoic glacial geology, 1871–1997. *Geological Society of London Memoir, 36*, 17–37.

Hoffman, P. F., and Schrag, D. P. (2002). The snowball Earth hypothesis: Testing the limits of global change. *Terra Nova, 14*, 129–155.

*Imbrie, J., and Imbrie, K. P. (1986). *Ice ages: Solving the mystery* (Rev. ed.). Harvard University Press.

Kopp, R. E., and others. (2005). The Paleoproterozoic snowball Earth: A climate disaster triggered by the evolution of oxygenic photosynthesis. *Proceedings of the National Academy of Sciences, 10*, 11131–11136.

Martin, P. E., and others. (2023). The rise of New Guinea and the fall of Neogene global temperatures. *Proceedings of the National Academy of Sciences, 120*, Article e2306492120.

Raymo, M. E., and Ruddiman, W. F. (1992). Tectonic forcing of late Cenozoic climate. *Nature, 359*, 117–122.

*Royer, D. L., and others. (2004). CO_2 as a primary driver of Phanerozoic climate. *GSA Today, 14*, 4–10.

Schmitz, O. (2018). Animals and the zoogeochemistry of the carbon cycle. *Science, 362*, eaar3213.

Steinthorsdottir, M., and others. (2021). The Miocene: The future of the past. *Paleoceanography and Paleoclimatology, 36*, Article e2020PA004037.

Van Groenigen, J. W., and others. (2014). Earthworms increase plant production: A meta-analysis. *Scientific Reports, 4*, Article 6365.

Walker, J.C.G., and others. (1981). A negative feedback mechanism for the long-term stabilization of Earth's surface temperature. *Journal of Geophysical Research Oceans, 86*, 9776–9782.

Zachos, J., and others. (2001). Trends, rhythms, and aberrations in global climate 65 Ma to present. *Science, 292*, 686–693.

Zahnle, K., and others. (2007). Emergence of a habitable planet. *Space Science Reviews, 129*, 35–78.

CHAPTER 15

Alvarez, L. W., and others. (1980). Extraterrestrial cause for the Cretaceous–Tertiary extinction. *Science, 208*, 1095–1108.

*Alvarez, W. (2015). *T. rex and the crater of doom* (Rev. ed.). Princeton University Press.

Bambach, R. K. (2006). Phanerozoic biodiversity: Mass extinctions. *Annual Review of Earth and Planetary Sciences, 34*, 127–155.

Bambach, R. K., and others. (2004). Origination, extinction, and mass depletions of marine diversity. *Paleobiology*, *30*, 522–542.

*Benton, M. (2023). *Extinctions: How life survives, adapts and evolves*. Thames and Hudson.

*Black, R. (2025). *When the Earth was green: Plants, animals, and evolution's greatest romance*. St. Martin's Press.

Caldeira, K., and Kasting, J. F. (1992). The life span of the biosphere revisited. *Nature*, *360*, 721–723.

Dal Corso, J., and others. (2020). Extinction and dawn of the modern world in the Carnian (Late Triassic). *Science Advances*, *6*, eaba0099.

Farnsworth, A., and others. (2023). Climate extremes likely to drive land mammal extinction during next supercontinent assembly. *Nature Geosciences*, *16*, 901–908.

Finnegan, S., and others. (2012). Climate change and the selective signature of the Late Ordovician mass extinction. *Proceedings of the National Academy of Sciences*, *109*, 6829–6834.

Gomes, M., and others. (2020). Taphonomy of biosignatures in microbial mats on Little Ambergris Cay, Turks and Caicos Islands. *Frontiers in Earth Science*, *8*, Article 576712.

Greene, S., and others. (2012). Recognising ocean acidification in deep time: An evaluation of the evidence for acidification across the Triassic–Jurassic boundary. *Earth-Science Reviews*, *113*, 72–93.

Hoffman, P. F. (2016). Cryoconite pans on Snowball Earth: Supraglacial oases for Cryogenian eukaryotes? *Geobiology*, *14*, 531–542.

Hoffman, P. F. (2025). Ecosystem relocation on Snowball Earth: Polar–alpine ancestry of the extant surface biosphere? *Proceedings of the National Academy of Sciences*, *122*, Article e2414059122.

Hoffman, P. F., and others. (2017). Snowball Earth climate dynamics and Cryogenian geology-geobiology. *Science Advances*, *3*, Article e1600983.

Hull, P., and others. (2020). On impact and volcanism across the Cretaceous–Paleogene boundary. *Science*, *367*, 266–272.

Knoll, A. H., and others. (2007). A paleophysiological perspective on the End-Permian mass extinction and its aftermath. *Earth and Planetary Science Letters*, *256*, 295–313.

Lovelock, J. E., and Whitfield, M. (1982). Life span of the biosphere. *Nature*, *296*, 561–563.

*Lynskey, D. (2024). *Everything must go: The stories we tell about the end of the world*. Picador.

Lyson, T. R. (2019). Exceptional continental record of biotic recovery after the Cretaceous–Paleogene mass extinction. *Science*, *366*, 977–983.

*Margulis, L., and Sagan, D. (1986). *Microcosmos: Four billion years of evolution from our microbial ancestors*. Summit Books.

Muscente, A. D., and others. (2018). Quantifying ecological impacts of mass extinctions with network analysis of fossil communities. *Proceedings of the National Academy of Sciences, 115*, 5217–5222.

Penn, J. L., and others. (2018). Temperature-dependent hypoxia explains biogeography and severity of end-Permian marine mass extinction. *Science, 362*, eaatl327.

Peters, S. (n.d.). *Sepkoski's online genus database*. https://strata.geology.wisc.edu/jack

Rasmussen, C.M.O., and others. (2023). Was the Late Ordovician mass extinction truly exceptional? *Trends in Ecology and Evolution, 38*, 812–821.

Raup, D. M., and Sepkoski, J. J. (1982). Mass extinctions in the marine fossil record. *Science, 215*, 1501–1503.

Schulte, P., and others. (2010). The Chicxulub asteroid impact and mass extinction at the Cretaceous–Paleogene boundary. *Science, 327*, 1214–1218.

Tamre, E., and Fournier, G. (2025). Recent origin of modern clades of iron oxidizers and low clade fidelity of iron metabolisms. *Applied and Environmental Microbiology*, 0.e01662-24.

CHAPTER 16

Bastin, J. P., and others. (2019). Understanding climate change from a global analysis of city analogues. *PLOS One, 14*, Article e0217592.

Bataille, C. (2025). Built to remove carbon. *Science, 387*, 134–135.

Blanchard-Wigglesworth, E., and others. (2025). Increasing boreal fires reduce future global warming and sea ice loss. *Proceedings of the National Academy of Sciences, 122*, Article e2424614122.

Christidis, N., and others. (2023). Rapidly increasing likelihood of exceeding 50 °C in parts of the Mediterranean and the Middle East due to human influence. *NPJ Climate and Atmospheric Science, 6*, Article 45.

*Crutzen, P. J., and Stoermer, E. F. (2000, May). The "Anthropocene." *Global Change Newsletter* 41, 17–18.

*Diaz, S., and others. (2019). Pervasive human-driven decline of life on Earth points to the need for transformative change. *Science, 366*, eaax3100.

Dirzo, R., and others. (2014). Defaunation in the Anthropocene. *Science, 345*, 401–406.

*Dressler, A. E. (2021). *Introduction to modern climate change* (3rd ed.). Cambridge University Press.

Elhacham, E., and others. (2020). Global human-made mass exceeds all living biomass. *Nature, 588*, 442–444.

Grant, P., and others. (2017). Evolutionary responses to extreme events. *Philosophical Transactions of the Royal Society B, 372*, Article 20160146.

*International Energy Agency. (2021). *Methane tracker 2021: Methane and climate change*. https://www.iea.org/reports/methane-tracker-2021/methane-and-climate-change

*Knoll, A. H. (2021). *A brief history of Earth: Four billion years in eight chapters*. HarperCollins.

*Kolbert, E. (2014). *The sixth extinction: An unnatural history*. Henry Holt and Company.

Kousky, C., Treuer, G., and Mache, K. J. (2024). Insurance and climate risks: Policy lessons from three bounding scenarios. *Proceedings of the National Academy of Sciences, 121*, Article e2317875121.

Lenton, T. M., and others. (2023). Quantifying the human cost of global warming. *Nature Sustainability, 6*, 1237–1247.

*Mann, M. E. (2023). *Our fragile moment: How lessons from Earth's past can help us survive the climate crisis*. PublicAffairs.

*Office of Energy Efficiency and Renewable Energy. (n.d.). *Renewable energy pillar*. https://www.energy.gov/eere/renewable-energy

*Phillips, R., and Milo, R. (n.d.). *The human impacts database*. http://www.anthroponumbers.org

*Popovich, N. (2023, November 20). How electricity is changing, country by country. *New York Times*. https://www.nytimes.com/interactive/2023/11/20/climate/global-power-electricity-fossil-fuels-coal.html

*Popovich, N. (2024, August 2). How does your state make electricity? *New York Times*. https://www.nytimes.com/interactive/2024/08/02/climate/electricity-generation-us-states.html

Pörtner, H. O., and others. (2004). Biological impact of elevated ocean CO_2 concentrations: Lessons from animal physiology and earth history. *Journal of Oceanography, 60*, 705–718.

*Pörtner, H. O., and others. (2022). *Climate change 2022: Impacts, adaptation and vulnerability: Summary for policymakers*. Intergovernmental Panel on Climate Change.

*Ripple, W. J., and others. (2024). The 2024 state of the climate report: Perilous times on planet Earth. *Bioscience, 74*, 812–824.

*Scripps Institution of Oceanography. (n.d.). *The Keeling Curve* [Atmospheric CO_2 levels, updated daily]. https://keelingcurve.ucsd.edu

Taing, A., and Kemp, L. (2021). A fate worse than warming? Stratospheric aerosol injection and global catastrophic risk. *Frontiers in Climate, 3*, Article 720312.

*Tollefson, J. (2024). Earth shattered heat records in 2023 and 2024: Is global warming speeding up? *Nature*, *637*, 523–524.

*Voosen, P. (2025). Earth's clouds are shrinking, boosting global warming. *Science*, *387*, 17–18.

*Zalasiewicz, J., Thomas, J. A., Waters, C. N., Turner, S., and Head, M. J. (2024). What should the Anthropocene mean? *Nature*, *362*, 980–984.

CHAPTER 17

Eigenbrode, J. L., and others. (2018). Organic matter preserved in 3-billion-year-old mudstones at Gale Crater. *Science*, *360*, 1096–1100.

Farley, K. A., and others. (2022). Aqueously altered igneous rocks sampled on the floor of Jezero crater, Mars. *Science*, *377*. https://doi.org/10.1126/science.abo219

Fernandez-Remolar, D. C., Morris, R., Gruener, J. E., Amils, R., and Knoll, A. H. (2005). The Rio Tinto basin: Mineralogy, sedimentary geobiology, and implications for interpretation of outcrop rocks at Meridiani Planum. *Earth and Planetary Science Letters*, *240*, 149–167.

Grotzinger, J. P., and others. (2005). Physical sedimentology of outcrop rocks at Meridiani Planum, Mars. *Earth and Planetary Science Letters*, *240*, 11–72.

Grotzinger, J. P., and others. (2014). A habitable fluvio-lacustrine environment at Yellowknife Bay, Gale Crater, Mars. *Science*, *343*, Article 1242777.

Horgan, B.H.N., and others. (2020). The mineral diversity of Jezero crater: Evidence for possible lacustrine carbonates on Mars. *Icarus*, *339*, Article 113526.

Hurowitz, J. A., and McLennan, S. M. (2007). A ~ 3.5 Ga record of water-limited, acidic weathering conditions on Mars. *Earth and Planetary Science Letters*, *260*, 432–443.

Hurowitz, J. A., and others. (2017). Redox stratification of an ancient lake in Gale crater, Mars. *Science*, *356*, eaah684.

Hurowitz, J. A., and others (2025). Redox-driven mineral and organic associations in Jezero Crater, Mars. *Nature*, *645*, 332–340.

*Jolliff, B., and others. (2019). Mars exploration rover Opportunity: Water and other volatiles on ancient Mars. In J. Filiberto and S. Schwenzer (Eds.), *Volatiles in the Martian crust* (pp. 285–328). Elsevier.

Kite, E. S., and Conway, S. (2024). Geological evidence for multiple climate transitions on early Mars. *Nature Geoscience*, *17*, 10–19.

*Knoll, A. H., and others. (2005). An astrobiological perspective on Meridiani Planum. *Earth and Planetary Science Letters*, *240*, 179–189.

*Lane, K.M.D. (2010). *Geographies of Mars: Seeing and knowing the Red Planet*. University of Chicago Press.

Lowell, P. (1906). *Mars and its canals*. Macmillan.

Mangold, N., and others. (2021). Perseverance rover reveals an ancient delta-lake system and flood deposits at Jezero crater, Mars. *Science, 374*, 711–719.

McKay, D., and others. (1996). Search for past life on Mars: Possible relic biogenic activity in Martian meteorite ALH84001. *Science, 273*, 924–930.

McLennan, S. M., and others. (2005). Provenance and diagenesis of the evaporite-bearing Burns formation, Meridiani Planum, Mars. *Earth and Planetary Science Letters, 240*, 95–121.

Milliken, R. E., and others. (2010). Paleoclimate of Mars as captured by the stratigraphic record in Gale Crater. *Geophysical Research Letters, 37*, Article L04201.

*National Aeronautics and Space Administration. (2024, July 25). *NASA's Perseverance rover scientists find intriguing Mars rock*. https://www.nasa.gov/missions/mars-2020-perseverance/perseverance-rover/nasas-perseverance-rover-scientists-find-intriguing-mars-rock

Squyres, S. W., and Knoll, A. H. (2005). Sedimentary rocks at Meridiani Planum: Origin, diagenesis, and implications for life on Mars. *Earth and Planetary Science Letters, 240*, 1–10.

Tosca, N. J., and Knoll, A. H. (2009). Juvenile chemical sediments and the long term persistence of water at the surface of Mars. *Earth and Planetary Science Letters, 286*, 379–386.

Tosca, N. J., and others. (2008). Water activity and the challenge for life on early Mars. *Science, 320*, 1204–1207.

Valantinas, A., and others. (2025). Detection of ferrihydrite in Martian red dust records ancient cold and wet conditions on Mars. *Nature Communications, 16*. https://doi.org/10.1038/s41467-025-56970-z

Wallace, A. R. (1907). *Is Mars habitable? A critical examination of Professor Percival Lowell's book "Mars and Its Canals," with an alternative explanation*. Macmillan.

Wordsworth, R., and others. (2021). An integrated scenario for the climate and redox evolution of Mars. *Nature Geoscience, 14*(3). https://doi.org/10.1038/s41561-021-00701-8

Wordsworth, W., and Coleridge, S. T. (1798). *Lyrical ballads, with a few other poems*. J. & A. Arch, Publishers.

CHAPTER 18

Bains, W. (2024). Source of phosphine on Venus—An unsolved problem. *Frontiers in Astronomy and Space Sciences, 11*, Article 1372057.

*Catling, D. L. (2014). *Astrobiology: A very short introduction.* Oxford University Press.

Charbonneau, D., and others. (2000). Detection of planetary transits across a sun-like star. *Astrophysical Journal, 529,* L45–L48.

*Clery, D. (2024). No place like home. *Science, 384,* 1286–1290.

Constantinou, T. N., Shorttle, O., and Rimmer, P. B. (2025). A dry Venusian interior constrained by atmospheric chemistry. *Nature Astronomy, 9,* 189–198.

Crawford, I. A., and Schulze-Makuch, D. (2024). Is the apparent absence of extraterrestrial technological civilizations down to the zoo hypothesis or nothing? *Nature Astronomy, 8,* 44–49.

Davila, A. F., and Eigenbrode, J. L. (2024). Enceladus: Astrobiology revisited. *Journal of Geophysical Research Biogeosciences, 129,* Article e2023JG007677.

De Kleer, K., and others. (2024). Isotopic evidence of long-lived volcanism on Io. *Science, 384,* 682–687.

Dumusque, X., and others. (2012). An Earth-mass planet orbiting α Centauri B. *Nature, 491,* 207–211.

Gillon, M., and others. (2017). Seven temperate terrestrial planets around the nearby ultracool dwarf star TRAPPIST-1. *Nature, 542,* 456–460.

*Green, J. (2023). *The possibility of life: The history and future of our search for life in the universe.* Hanover Square Press.

*Hand, K. (2020). *Alien oceans: The search for life in the depths of space.* Princeton University Press.

Hörst, S. M. (2017). Titan's atmosphere and climate. *Journal of Geophysical Research Planets, 122,* 432–482.

*Howell, E. (2023). The Fermi Paradox: Where are the aliens? https://www.space.com/25325-fermi-paradox.html

*Kaltenegger, L. (2024). *Alien earths: The new science of planet hunting in the cosmos.* St. Martin's Press.

*Kasting, J. F. (2012). *How to find a habitable planet.* Princeton University Press.

Kasting, J. F., and others. (1993). Habitable zones around main-sequence stars. *Icarus, 101,* 108–128.

*Lingam, M., and Balbi, A. (2024). *From stars to life: A quantitative approach to astrobiology.* Cambridge University Press.

Lingam, M., and Loeb, A. (2019). Relative likelihood of success in the search for primitive versus intelligent extraterrestrial life. *Astrobiology, 19,* 28–39.

Mastrogiuseppe, M., and others. (2019). Deep and methane-rich lakes on Titan. *Nature Astronomy, 3,* 535–542.

Mayor, M., and Queloz, D. (1995). A Jupiter-mass companion to a solar-type star. *Nature, 378,* 353–359.

*National Aeronautics and Space Administration. (n.d.). *Exoplanets.* https://science.nasa.gov/exoplanets

Niemann, H. B., and others. (2005). The abundances of constituents of Titan's atmosphere from the GCMS instrument on the Huygens probe. *Nature, 438,* 779–784.

O'Callaghan, J. (2025, March 11). Saturn gains 128 new moons, bringing its total to 274. *New York Times.* https://www.nytimes.com/2025/03/11/science/saturn-new-moons.html

Pappalardo, R. T., and others. (2024). Science overview of the Europa Clipper mission. *Space Science Reviews, 220,* Article 4. https://doi.org/10.1007/s11214-024-01070-5

Peters, J., and others. (2024). Detection of HCN and diverse redox chemistry in the plume of Enceladus. *Nature Astronomy, 8,* 164–173.

Porco, C. C., and others. (2006). Cassini observes the active south pole of Enceladus. *Science, 311,* 1383–1401.

Postberg, F., and others. (2011). A salt-water reservoir as the source of a compositionally stratified plume on Enceladus. *Nature, 474,* 620–622.

Postberg, F., and others. (2018). Macromolecular organic compounds from the depths of Enceladus. *Nature, 558,* 564–568.

Reinhard, C. T., Olson, S. L., Schwieterman, E. W., and Lyons, T. W. (2017). False negatives for remote life detection. *Astrobiology, 17,* 287–297.

Roberts, J. H., and Nimmo, F. (2008). Tidal heating and the long-term stability of a subsurface ocean on Enceladus. *Icarus, 194,* 675–689.

*Scoles, S. (2017). *Making contact: Jill Tarter and the search for extraterrestrial intelligence.* Pegasus Books.

Seager, S. (2013). Exoplanet habitability. *Science, 340,* 577–581.

*Shostak, S. (2009). *Confessions of an alien hunter: A scientist's search for extraterrestrial intelligence.* National Geographic.

Stofan, E. R., and others. (2007). The lakes of Titan. *Nature, 445,* 61–64.

Tomasko, M. G., and others. (2005). Rain, winds and haze during the Huygens probe's descent to Titan's surface. *Nature, 438,* 765–778.

Vickers, P. (2025). Surveys of the scientific community on the existence of extraterrestrial life. *Nature Astronomy, 9,* 16–18.

Waite, J. H., and others. (2017). Cassini finds molecular hydrogen in the Enceladus plume: Evidence for hydrothermal processes. *Science, 356,* 155–159.

ILLUSTRATION CREDITS

FIGURE 2.1. Author, various sources, including Jet Propulsion Laboratory, Global Carbon Cycle: https://airs.jpl.nasa.gov/resources/155/global-carbon-cycle/

FIGURE 3.1. Adapted from *Biology: How Life Works* (4th ed.) by James Morris, and others. Copyright 2023, 2019, 2016, 2013 by Macmillan Learning. All rights reserved. Reprinted with permission of Macmillan Learning

FIGURE 3.2. Data from G. Filippelli, 2002, *Reviews in Mineralogy and Geochemistry* 48: 391–425

FIGURE 3.3. NASA Earth Observatory, NASA Earth Observatory, https://earthobservatory.nasa.gov/images/4097/global-chlorophyll

FIGURE 4.1. Data from Ask a Biologist, Arizona State University: https://askabiologist.asu.edu/content/atoms-life

FIGURE 4.2. Adapted from *Biology: How Life Works* (4th ed.) by James Morris, and others. Copyright 2023, 2019, 2016, 2013 by Macmillan Learning. All rights reserved. Reprinted with permission of Macmillan Learning

FIGURE 6.1. Wikimedia Commons, public domain, via Wikimedia Commons

FIGURE 6.2. Data from Pozzi, L., and others, 2014, *Molecular Phylogenetics and Evolution* 75: 165–183

FIGURE 6.3. Multiple sources, constructed from a number of sources

FIGURE 8.1. Photo: A. Knoll

FIGURE 9.1. Photo: A. Knoll

FIGURE 9.2. Photo: A. Knoll

FIGURE 9.3. Adapted from Olejarz and others (2021).

FIGURE 10.1. From Yanan Shen, courtesy of Yanan Shen

FIGURE 10.2 A and B. Photo: A. Knoll

FIGURE 10.2C. From Maoyan Zhu and Lanyon Miao, courtesy of Lanyun Miao and Maoyan Zhu

FIGURE 11.1. Photo: A. Knoll

FIGURE 11.2. Photo: A. Knoll

FIGURE 11.3. Photo: A. Knoll

FIGURE 12.1. Photo: A. Knoll

FIGURE 12.2. A: Wikimedia Commons, photo by Andreas Drews; B: Wikimedia Commons, photo by Berezovska

FIGURE 12.3. Photo: A. Knoll

FIGURE 12.4. From Phoebe Cohen, courtesy of Phoebe Cohen

FIGURE 12.5A. From Shuhai Xiao, courtesy of Shuhai Xiao

FIGURE 12.5B. Photo: A. Knoll

FIGURE 12.6. Photo: A. Knoll

FIGURE 13.1. Photo: A. Knoll

FIGURE 14.1. Photo: A. Knoll

FIGURE 14.2. Photo: A. Knoll

FIGURE 15.1. Photo: A. Knoll

FIGURE 15.2. https://strata.geology.wisc.edu/jack/, constructed by Shanan Peters from the database pioneered by Jack Sepkoski. https://strata.geology.wisc.edu/jack/

FIGURE 16.1. https://www.bbc.com/news/articles/cd7575x8yq5o, BBC: https://www.bbc.com/news/articles/cd7575x8yq5o

FIGURE 16.2. Environmental Protection Agency, Environmental Protection Agency: https://www.epa.gov/ghgemissions/sources-greenhouse-gas-emissions

FIGURE 17.1. NASA/JPL-Caltech/Arizona State University

FIGURE 17.2. D. Savransky and J. Bell/JPL/NASA/Cornell/ASUNASA/JPL

FIGURE 17.3. A: NASA/JPL; B: A. Knoll

FIGURE 17.4. NASA/JPL-Caltech, ASU

FIGURE 18.1. NASA/JPL-Caltech/DLR

FIGURE 18.2. NSA/JPL/Space Science Institute

FIGURE 18.3. ESA/NASA/JPL/University of Arizona

FIGURE 18.4. https://www.esa.int/Science_Exploration/Space_Science/How_to_find_an_extrasolar_planet, European Space Agency's How to find an extrasolar planet webpage

INDEX

Page numbers in italics refer to figures and tables.